# NAVIGATING
# METABOLISM

**ALSO FROM COLD SPRING HARBOR LABORATORY PRESS**

*Cell Survival and Cell Death*
*Mitochondria*
*MYC and the Pathway to Cancer*
*Protein Homeostasis*
*Quickstart Molecular Biology*
*Signaling by Receptor Tyrosine Kinases*
*Signal Transduction*
*Type 1 Diabetes*
*Yeast Intermediary Metabolism*

# NAVIGATING METABOLISM

---

NAVDEEP S. CHANDEL

*Northwestern University*
*Feinberg School of Medicine*

*Illustrations by Pete Jeffs*

**COLD SPRING HARBOR LABORATORY PRESS**
Cold Spring Harbor, New York • www.cshlpress.org

**NAVIGATING METABOLISM**

| | |
|---|---|
| **Publisher** | John Inglis |
| **Acquisition Editor** | Richard Sever |
| **Director of Editorial Development** | Jan Argentine |
| **Developmental Editor** | Judy Cuddihy |
| **Project Manager** | Inez Sialiano |
| **Permissions Coordinator** | Carol Brown |
| **Production Editor** | Kathleen Bubbeo |
| **Production Manager** | Denise Weiss |
| **Cover Designer** | Pete Jeffs |

*Front cover artwork:* This rendering of a cell illustrates the metabolism of glucose, fatty acids, and amino acids through glycolysis and mitochondria and symbolizes this book's exploration of how anabolic and catabolic pathways are used for bioenergetics, biosynthesis, and signal transduction.

Library of Congress Cataloging-in-Publication Data

Chandel, Navdeep S. (Navdeep Singh)
Navigating metabolism / Navdeep Chandel, Northwestern Medical School; Illustrations by Peter Jeffs

      pages cm

Summary: "Metabolic pathways used to be "road maps" most biologists learned as undergraduates and then promptly forgot. Recent work has revealed how changes in metabolism are closely linked to many aspects of cell behavior and the development of cancer and other diseases. This book represents both a new look at metabolism and a refresher course. It surveys the major metabolic pathways, places these in biological context, and highlights the key control points that control cell behavior and can become dysregulated in disease."-- Provided by publisher.

  ISBN 978-1-62182-040-6 (cloth)- -ISBN 978-1-62182-129-8 (paper)
1.   Cell metabolism. I. Title.

  QH634.5.C53 2015
  572'.4--dc23

                           2014029307

For a complete catalog of all Cold Spring Harbor Laboratory Press publications, visit our website at www.cshlpress.org.

*I dedicate this book to the previous and present members of my laboratory, who inspired me to write it. They provided the initial ideas for the tone and topics and helped edit the final proofs.*

# Contents

# Common Abbreviations

These common abbreviations are used without definition throughout the book.

| | |
|---|---|
| acetyl-CoA | acetyl coenzyme A |
| AMP | adenosine 5′-monophosphate |
| ADP | adenosine 5′-diphosphate |
| ATP | adenosine 5′-triphosphate |
| ATPase | adenosine triphosphatase |
| CMP | cytidine 5′-monophosphate |
| CDP | cytidine 5′-diphosphate |
| CTP | cytidine 5′-triphosphate |
| $CO_2$ | carbon dioxide |
| CoA | coenzyme A |
| FAD | flavin adenine dinucleotide |
| $FADH_2$ | reduced form of flavin adenine dinucleotide |
| GMP | guanosine 5′-monophosphate |
| GDP | guanosine 5′-diphosphate |
| GSH | glutathione |
| GSSH | oxidized form of glutathione |
| GTP | guanosine 5′-triphosphate |
| $H_2O_2$ | hydrogen peroxide |
| $NAD^+$ | nicotinamide adenine dinucleotide phosphate |
| NADH | reduced form of nicotinamide adenine dinucleotide phosphate |
| $NADP^+$ | nicotinamide adenine dinucleotide phosphate |
| NADPH | reduced form of nicotinamide adenine dinucleotide phosphate |
| $O_2^-$ | superoxide |
| ROS | reactive oxygen species |
| SOD | superoxide dismutase |
| TCA | tricarboxylic acid |
| TDP | thymidine 5′-diphosphate |
| TMP | thymidine 5′-monophosphate |

| | |
|---|---|
| TTP | thymidine 5′-triphosphate |
| UCP | uncoupling protein of the mitochondrial inner membrane |
| UDP | uridine 5′-diphosphate |
| UMP | uridine 5′-monophosphate |
| UTP | uridine 5′-triphosphate |

# Foreword

IN THE 20TH CENTURY, STUDIES OF cellular and organismal metabolism led to the delineation of how nutrients were catabolized intracellularly to produce life-sustaining levels of ATP by the combined activities of enzymes involved in intermediate metabolism and the citric acid cycle. This golden age of metabolism ended with the crowning discovery of mitochondrial oxidative phosphorylation by Peter Mitchell in the 1960s. The elucidation of ATP generation was one of the greatest collective achievements of modern science and remains until this day the foundation of all biochemistry textbooks written for high school, college, and graduate students. However, by 1978, when Mitchell was awarded his Nobel Prize, biologists armed with the tools of molecular biology had moved on to studying gene expression, signal transduction, cell division, and differentiation. The fact that these processes were all anabolic and resource-intensive was largely ignored.

It was only at the start of the 21st century that it became apparent that the previous century's scientific stars had not quite worked out how metabolism was reprogrammed to support net macromolecular synthesis and cellular/organismal growth. Although individual scientists have been recognized for elucidating how nucleic acids, lipids, and amino acids were synthesized, the pathways they elucidated were viewed as homeostatically regulated branches of traditional intermediate metabolism.

During the last 15 years, the view that nutrient uptake and metabolism is regulated through cell-autonomous homeostatic mechanisms has been challenged by the discovery that signal transduction can directly influence nutrient uptake. For example, receptor-activated kinases were shown to directly induce nutrient uptake and metabolism. Furthermore, subsequent studies of the impact of nutrient uptake and availability on gene expression have demonstrated that the levels of intermediate metabolites can in turn regulate chromatin organization, gene expression, and protein translation. In turn, there is also a growing appreciation that signal transduction–initiated, transcriptionally induced variations of enzyme expression levels can influence intracellular metabolite flux into anabolic versus catabolic fates.

The integration of the role of anabolic metabolism in cell signaling and gene expression is still in its infancy. In *Navigating Metabolism*, Navdeep Chandel provides the reader with the conceptual framework through which to understand the evolving concepts that are reshaping and revitalizing metabolite biochemistry. Chandel brings a fresh perspective to metabolism, as he is equally comfortable and conversant in traditional biochemistry, bioenergetics, molecular biology, and signal transduction. Furthermore, the book provides readers with compelling examples of how Chandel and other scientists (Chapter 12) have drawn from these fields to reexamine how alterations in cellular/organismal metabolism contribute to the pathogenesis of diseases, including diabetes, cancer, and neurodegeneration. *Navigating Metabolism* is not meant to replace existing biochemistry textbooks, but rather to enhance and provide an update for all who are interested in the emerging concept of how metabolism is integrated with the rest of modern biology.

CRAIG B. THOMPSON
*Memorial Sloan Kettering Cancer Center*

# Preface

Today the study of metabolism is enjoying a renaissance. It might not be evident to some of the younger readers but not so long ago metabolism was relegated to the background of scientific endeavors. How did metabolism make a comeback? I can only offer my own personal journey as an illustration of what I think excited many of the scientists to reengage with metabolism. In 1989, I was in the midst of my undergraduate student studies in mathematics at the University of Chicago. During the summer, I joined a biology laboratory to pay the bills so that I could spend the summer in Chicago. My project was to examine whether 3-phosphoglycerate could prevent hypoxia (low oxygen)-induced cell death. As a mathematician, I found that metabolism had a great appeal to me because metabolic pathways have an inherent logic dictated by thermodynamics. I continued working on metabolism in Paul Schumacker's laboratory as a graduate student, and my thesis examined the effects of hypoxia on the kinetics of the mitochondrial enzyme cytochrome *c* oxidase.

By the mid-1990s it was clear to me that most of my colleagues had very little interest in metabolism or any of my experiments. Most graduate students and postdocs at that time were excited about genetics, molecular biology, and signal transduction. But soon I was exposed to two ideas that for me connected metabolism to the rest of biology. First, while playing soccer, I met Craig Thompson, and he told me about an exciting finding from Xiaodong Wang's laboratory regarding the release of cytochrome *c* from mitochondria as an essential step in activating caspases resulting in death of mammalian cells. Craig's laboratory already had been studying metabolism in the context of growth factor regulation of apoptosis, and now his laboratory was further invigorated by Xiaodong's findings. My initial response was that the release of cytochrome *c* from mitochondria was an artifact. It was hard for me to imagine that cytochrome *c* was doing anything else other than being a substrate for cytochrome *c* oxidase. I just could not imagine that cytochrome *c* had another function outside the mitochondria, as the key signal to execute cell death. Of course, Craig had already seen the broad implications and connections of metabolism to biology beyond apoptosis. Soon, I was immersed in Craig's world of connecting metabolism to biology, primarily by working with a talented graduate student in his laboratory, Matthew Vander Heiden. Matt explained the details of apoptosis to me, and soon we were engaged in daily conversations about the cross talk between apoptosis and metabolism, which in retrospect was instrumental in shaping my views on metabolism—not as an isolated discipline but connected to fundamental cellular decisions like "whether to live or die."

The second idea that brought my myopic vision of metabolism from biochemical pathways to metabolisms's vivid interaction with the rest of biology was an introduction to hypoxic gene expression by the late Eugene Goldwasser. Luckily, Emin Maltepe, a soccer buddy and graduate student in Celeste Simon's laboratory, was engaged in the first knockout of the transcription factor hypoxia-inducible factor (HIF), which is essential for induction of hypoxic genes. Emin explained the importance of HIF to me, and it was not very long before Paul, Emin, Celeste, Eugene, and I were all engaged in linking mitochondria to HIF. The idea that mitochondria could regulate gene expression further solidified my interest in pursuing the connection of metabolism to the rest of cell biology. When I had finished my postdoc in late 1999 and embarked in setting up my own laboratory in January of 2000, it was clear to me that going forward the excitement about metabolism revolved around integrating it to the rest of biology, physiology, and disease. Mitochondria were now more than bioenergetic and biosynthetic organelles—they were signaling organelles! Thus, the cross talk of metabolism to biology has reenergized this "ancient" discipline.

In the past decade, examining metabolism and its links to many biological outputs has intrigued many seasoned investigators along with new generations of students and postdocs. But most have found revisiting their biochemistry books quite intimidating. I flirted with the idea of writing a simple introduction to metabolism book with a conceptual framework of metabolic pathways and their links to the rest of biology, a book that would not replace many of the spectacular existing biochemistry textbooks but would complement them by illustrating how biochemistry comes alive through metabolism's connection to biology, physiology, and disease. However, one of the edicts of being a research scientist is not to engage in writing a single-author book. The daily pressure to get grants, publish, and do innovative work is challenging enough. So it was just my luck that I tagged along as an uninvited guest with Doug Green and met Richard Sever, Assistant Director and Acquisition Editor of Cold Spring Harbor Laboratory Press. Richard had been contemplating a similar idea, and before long I was writing my first book! It was quite challenging to get started and stay the course to the finish line. Luckily, I had Judy Cuddihy as an editor. Judy brought her vast experience in publishing to shape this book. Importantly, I am very grateful to her because throughout this process she has been both a coach and fan cheering me to the finish line. The artwork in this book is vital to making metabolism engaging, and I am appreciative to Pete Jeffs for his spectacular figures throughout the book. *Navigating Metabolism* should be viewed as a wonderful collaboration between Judy, Pete, and me. I am also grateful to the staff at Cold Spring Harbor Laboratory Press, especially Inez Sialiano, Project Manager, and Kathleen Bubbeo, Senior Production Editor. I realize some might be disappointed that I did not cover certain topics nor provide enough detail, but the objective here was to provide an accessible introductory book to reengage a broad range of investigators with metabolism.

Finally, there are a few people who have contributed immensely to creating this book. My friend and colleague Ralph DeBerardinis wrote a fantastic appendix on metabolomics for the book and has been very supportive throughout the past year. The present and past members of my laboratory have provided wonderful ideas and helped edit the proofs. I am thankful to David Sabatini for insisting my original titles for the book were not very engaging and to Gerard Evan for providing the clever title *Navigating Metabolism*. And I am indebted to the 11 authors who contributed to Chapter 12, "Future Pathways of Metabolism Research." Chapter 12 provides a sense that there is still much to be learned about how metabolism connects to biology. Most of all, I am thankful to my daughter Anjali and my wife Evangelina for their love and support through this enduring process. Anjali's daily excitement about the book put pressure on me to finish, and Evangelina has been patiently hearing me say over and over "the book will be done next month." Much love to both. Now the book is finally done, and hopefully its readers and my colleagues, friends, and family will enjoy it!

NAVDEEP S. CHANDEL
*August, 2014*

# 1

# Introduction to Metabolism

## WHAT IS METABOLISM?

I F YOU ASK MY YOUNG DAUGHTER, Anjali, she will tell you that metabolism is about how we break down food and build muscles. At first glance, this is not far from the definition of metabolism, which is a set of pathways to build and break down macromolecules that comprise living matter. These macromolecules include proteins, lipids, nucleic acids (DNA and RNA), and carbohydrates; their biosynthesis is referred to as anabolism. In contrast, these macromolecules can also be broken down into their constituents—amino acids, fatty acids, nucleotides, and sugars—in a process referred to as catabolism. Catabolism can generate energy in the form of ATP and reducing equivalents (NADH, NADPH, and $FADH_2$) to sustain chemical reactions of living organisms (Fig. 1-1). In contrast, anabolic reactions use ATP and reducing equivalents to reassemble products of catabolism. It is astonishing how the basic macromolecules of living matter and the metabolic pathways to generate energy and synthesize cell constituents are highly conserved among organisms. In fact, this profound similarity of basic metabolic pathways across life might be the best evidence for evolution, suggesting that all living organisms evolved from a common source.

## WHY STUDY METABOLISM?

Much of biology during the early part of the 20th century focused on elucidating anabolic and catabolic pathways, and, until the 1960s, many Nobel Prize awards in Physiology or Medicine and in Chemistry were given to these metabolism-related discoveries (see Box 1-1). After the discovery of DNA and protein phosphorylation in the mid-20th century, biology entered a new era of understanding gene regulation and signaling pathways. Metabolism was thought to change only as a consequence of changes in signaling pathways and gene expression. In other words, the phenotype of a cell or an organism feeds back on metabolism to acquire the necessary metabolites and energy to function. In the past two decades, however, accumulating evidence suggests that metabolism can actually regulate signaling pathways and gene expression, thereby playing a causal role in dictating divergent biological outcomes, such as

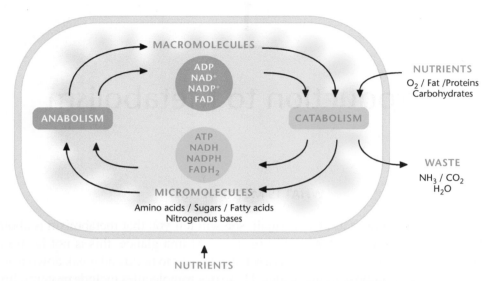

**Figure 1-1.** Metabolism is the flux through anabolic and catabolic pathways.

cell proliferation, cell death, differentiation, gene expression, and adaptation to stress. Hence, metabolism dictates phenotype. Moreover, an exciting idea is that changes in metabolism underlie diseases such as diabetes, neurodegeneration, cancer, hepatotoxicity, and cardiovascular and inflammatory diseases (Fig. 1-2). The use of widely adopted statins to interfere with cholesterol metabolism to diminish cardiovascular disease is an excellent example of the importance of understanding metabolism to prevent diseases (see Chapter 7). Other widely used medications, such as aspirin, an anti-inflammatory drug, and metformin, an antidiabetic drug, have recently been proposed to target AMP-activated protein kinase (AMPK) and mitochondrial complex I, respectively. It is with great hope that many scientists are testing whether intervening in metabolic pathways could alleviate the suffering caused by these diseases.

An emerging subject is metabolism's connection to signaling and induction of gene expression (see Chapter 10). For example, protein acetylation of lysine residues and protein oxidation of cysteine residues are important regulators of cellular signaling pathways. Acetylation requires acetyl-CoA as a substrate, and protein oxidation requires ROS. Metabolism controls the availability of both acetyl-CoA and ROS. Conversely, signaling pathways regulate metabolism. Notably, the kinase mammalian target of rapamycin (mTOR), which controls growth of cells and organs, as well cancer and the aging process, and thus is a major therapeutic target, exerts its major biological effects, in part, by balancing the anabolic and catabolic pathways within a cell based on nutrient availability. Until the past decade, it was thought that the induction of genes was restricted to regulation by transcription factors. However, in the past two decades, epigenetics has emerged as a powerful mechanism for controlling gene expression, and metabolism plays a role in this. For example,

## BOX 1-1. NOTABLE METABOLISM-RELATED NOBEL PRIZES

| Year | Name | Citation |
|------|------|----------|
| **Chemistry** | | |
| 1902 | Hermann Emil Fischer | "[for] his work on sugar and purine syntheses" |
| 1907 | Eduard Buchner | "for his biochemical researches and his discovery of cell-free fermentation" |
| 1927 | Heinrich Otto Wieland | "for his investigations of the constitution of the bile acids and related substances" |
| 1928 | Adolf Otto Reinhold Windaus | "for the services rendered through his research into the constitution of the sterols and their connection with the vitamins" |
| 1929 | Arthur Harden, Hans Karl August Simon von Euler-Chelpin | "for their investigations on the fermentation of sugar and fermentative enzymes" |
| 1937 | Walter Norman Haworth | "for his investigations on carbohydrates and vitamin C" |
| | Paul Karrer | "for his investigations on carotenoids, flavins and vitamins A and B2" |
| 1938 | Richard Kuhn | "for his work on carotenoids and vitamins" |
| 1970 | Luis F. Leloir | "for his discovery of sugar nucleotides and their role in the biosynthesis of carbohydrates" |
| 1978 | Peter D. Mitchell | "for his contribution to the understanding of biological energy transfer through the formulation of the chemiosmotic theory" |
| 1997 | Paul D. Boyer, John E. Walker | "for their elucidation of the enzymatic mechanism underlying the synthesis of ATP" |
| **Physiology or Medicine** | | |
| 1922 | Otto Fritz Meyerhof | "for his discovery of the fixed relationship between the consumption of oxygen and the metabolism of lactic acid in the muscle" |
| | Archibald Vivian Hill | "for his discovery relating to the production of heat in the muscle" |
| 1923 | Frederick Grant Banting, John James Rickard Macleod | "for the discovery of insulin" |
| 1929 | Christiaan Eijkman, | "for his discovery of the antineuritic vitamin" |
| | Sir Frederick Gowland Hopkins | "for his discovery of the growth-stimulating vitamins" |
| 1931 | Otto Heinrich Warburg | "for his discovery of the nature and mode of action of the respiratory enzyme" |
| 1937 | Albert Szent-Györgyi von Nagyrapolt | "for his discoveries in connection with the biological combustion processes, with special reference to vitamin C and the catalysis of fumaric acid" |
| 1947 | Carl Ferdinand Cori, Gerty Theresa Cori, née Radnitz | "for their discovery of the course of the catalytic conversion of glycogen" |
| | Bernardo Alberto Houssay | "for his discovery of the part played by the hormone of the anterior pituitary lobe in the metabolism of sugar" |

*(Continued)*

**BOX 1-1.** *(Continued)*

| Year | Name | Citation |
|------|------|----------|
| 1953 | Sir Hans Adolf Krebs | "for his discovery of the citric acid cycle" |
|      | Fritz Albert Lipmann | "for his discovery of coenzyme A and its importance for intermediary metabolism" |
| 1955 | Axel Hugo Theodor Theorell | "for his discoveries concerning the nature and mode of action of oxidation enzymes" |
| 1964 | Konrad Bloch, Feodor Lynen | "for their discoveries concerning the mechanism and regulation of the cholesterol and fatty acid metabolism" |
| 1985 | Michael S. Brown, Joseph L. Goldstein | "for their discoveries concerning the regulation of cholesterol metabolism" |

methylation of histones and DNA requires methyl groups, which are donated by methionine through one-carbon metabolic pathways. Furthermore, the enzymes that cause demethylation are members of the growing α-ketoglutarate-dependent dioxygenase family.

Cancer metabolism has generated a lot of excitement in the past few years (see Chapter 11). In 2008–2009, mutations in the metabolic enzymes isocitrate dehydrogenase (IDH) 1 and 2 in several types of malignant gliomas have linked the cancer genetics field to cancer metabolism. Surprisingly, these mutations do not cause major

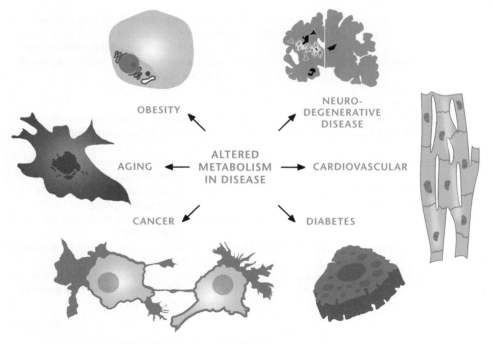

**Figure 1-2.** Altered metabolism is linked to many common diseases.

changes in metabolic pathways. Instead, the IDH mutations generate a new oncometabolite, 2-hydroxyglutarate (2HG), which regulates epigenetics by inhibiting demethylase enzymes that are members of the α-ketoglutarate-dependent dioxygenase family. Normally, 2HG is detectable at low levels in cells, but shows an increase of >100-fold in tumor samples with IDH mutations. These mutations have now been found to be prevalent in chondrosarcomas and acute myeloid leukemias (AMLs). Remarkably, today there are drugs in clinical trials that can distinguish between wild-type and mutant IDH and are being used for treatment in cancer patients with IDH mutations. Beyond mutations in metabolic enzymes, large-scale functional genetic screens have identified wild-type metabolic enzymes that are necessary for growth of certain types of cancers. Furthermore, large-scale bioinformatic analysis has deciphered a list of wild-type metabolic enzymes that are highly up-regulated across many cancer cells, thus potentially yielding new targets for cancer therapy.

Finally, "textbook" biochemistry has also been invigorated by discovery of new metabolic pathways and metabolic proteins, as well as novel modes of regulating metabolic pathways. For example, pyruvate kinases are one of the major enzymes that regulate glycolysis (see Chapter 3). Yet, the mechanisms by which proliferating cells modulate the activity of pyruvate kinases to control glycolysis to balance their biosynthetic and bioenergetics demand is just being unraveled. Moreover, although it been known for decades that pyruvate is metabolized by mitochondria, it is only recently that the transporter responsible for pyruvate transport into the mitochondria was discovered.

The use of isotope carbon-labeling techniques (see the Appendix) has elucidated interesting new pathways both in vitro and in vivo. For example, the metabolism of the amino acid glutamine through the TCA cycle has been well known for decades. However, in the past few years, through the use of isotope carbon-labeling techniques, it was discovered that glutamine metabolism not only fluxes through the canonical clockwise TCA cycle but also fluxes through a few reverse steps of the TCA cycle. New parameters governing metabolic flux through a pathway are also being revealed using metabolomics and computational modeling. And mitochondria, the central hub of metabolism, have undergone a makeover from being known for decades as the "powerhouse" of the cells to also being appreciated as "signaling organelles" because of their involvement in controlling many divergent biological processes, ranging from cell proliferation to cellular differentiation. The ensuing decades will, no doubt, yield more insight into how metabolism integrates with multiple different biological processes within cells.

## HOW SHOULD WE THINK ABOUT METABOLISM?

First, most of us think the importance of metabolic pathways lies in understanding how organisms generate sufficient amounts of energy to conduct biological func-

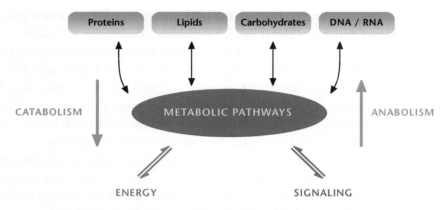

**Figure 1-3.** Overview of metabolic pathways covered in this book.

tions. However, an equally important function of metabolic pathways is to generate intermediates for biosynthesis of macromolecules (Fig. 1-3). As Hans Krebs, who received the Nobel Prize in Physiology or Medicine in 1953 for his discovery of the citric acid cycle, stated in his Nobel lecture, "Many observations, especially from isotope experiments, support the view that in some micro-organisms the cycle primarily supplies intermediates rather than energy, whilst in the animal and most other organisms it supplies both energy and intermediates." Throughout this book, we will pay attention to both the energy generation and biosynthesis aspects of metabolic pathways. There are illustrative metabolic pathways throughout the book as quick reference guides.

Second, it is essential to always examine why a particular reaction in a metabolic pathway goes in one direction versus another. There is a logic that dictates the directionality and rate of reactions within a given metabolic pathway. Thus, before elucidating metabolic pathways in Chapter 3, essential concepts, such as thermodynamic constraints, enzyme activity, and levels of products versus reactants that are key determinants of directionality and rates of any given reaction within a metabolic pathway, are discussed in Chapter 2. Although new metabolic pathways will be uncovered in the future, they are likely to be constrained by the same parameters outlined in Chapter 2.

Third, it is vital to think consistently about metabolic pathways in the context of biological and physiological functions. Metabolic pathways in cells are continuously responding to diverse stimuli, ranging from growth factors to changes in nutrients, as well as metabolic pathways, and are key decision makers in dictating biological outcomes. It makes sense that, before a cell commits to any biological function, there would be modes of communication to assess whether there is sufficient energy and biosynthetic activity. Throughout the book, metabolic pathways are put in the context of biological and physiological functions.

## HOW IS THIS BOOK ORGANIZED?

*Navigating Metabolism* is not intended to substitute for many of the excellent biochemistry textbooks already available. *Merriam-Webster's Dictionary* defines biochemistry as "chemistry that deals with the chemical compounds and processes occurring in organisms." For many of us, biochemistry looks like static metabolites pinned down in metabolic pathways. Metabolism brings those metabolic pathways to life, because it is about the flux through those pathways and how nutrient availability, genes, and signaling control metabolic pathways. Thus, Chapter 2 begins with parameters that control flux through metabolic pathways, followed by Chapters 3 and 4 on glycolysis and mitochondrial metabolism, the two central carbon metabolic pathways. Chapter 5 explores NADPH, which I refer to as "the forgotten reducing equivalent," as most of us only remember NADH, which is necessary to generate ATP. NADPH is necessary for driving many of the anabolic reactions in our cells. Chapters 6, 7, 8, and 9 describe metabolism relating to sugars, lipids, amino acids, and nucleotides, respectively. These four chapters have similar themes in which I describe the catabolic, anabolic, and signaling pathways for these building blocks. Chapter 10 discusses the cellular signaling pathways that control metabolic pathways. Chapter 11 uses metabolism of proliferating cells as an example to synthesize the concepts presented in the preceding 10 chapters. Finally, Chapter 12, entitled "The Future of Metabolism," is a collection of short essays written by scientists who are currently heavily engaged in investigating cellular metabolism.

# 2

# Basics of Metabolism

METABOLISM PERFORMS four essential functions for cells.

1. It provides energy by generating ATP to conduct cellular functions.

2. It converts nutrients, such as fat, protein, and sugar, into simpler structures, such as fatty acids, amino acids, and glucose, respectively (i.e., catabolism). This process can generate energy.

3. It converts simpler structures into macromolecules, such as nucleotides, lipids, and proteins (i.e., anabolism). This process requires energy.

4. It participates in cellular functions beyond energy, anabolism, and catabolism, such as cellular signaling and gene transcription. For example, metabolites serve as substrates for posttranslational modification of proteins to elicit changes in protein function or regulate epigenetics to invoke changes in gene expression.

As an example, the metabolism of acetyl-CoA fulfills all four of these functions (Fig. 2-1). Nutrients, such as fat, protein, and table sugar, are converted into simpler structures, such as fatty acids, amino acids, and glucose, respectively (i.e., catabolism). All three can generate acetyl-CoA, which can produce ATP in the mitochondria through the TCA cycle, also known as the citric acid cycle, or the Krebs cycle (i.e., energy). Acetyl-CoA can also be used to generate cholesterol, triacylglycerol, and phospholipids (i.e., anabolism). Acetyl-CoA is also a substrate for protein acetylation. Recent evidence indicates that protein acetylation is a robust posttranslational modification that alters protein function to change biological processes (i.e., signaling).

Metabolic pathways are a series of connected reactions that can be linear, circular, or bifurcate into two directions. These pathways are found in different compartments of the cell, such as the mitochondria, endoplasmic reticulum, and cytosol, and these reactions use metabolites, enzymes, and energy. The small molecules that are the reactants and products of a reaction are called metabolites, and the catalysts that drive the reactions by converting reactants into products are enzymes. Reactions can either require (absorb energy) or release energy. The levels of metabolites and

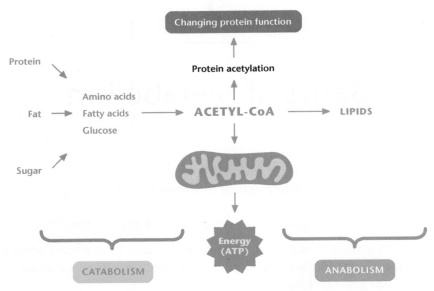

**Figure 2-1.** Acetyl-CoA fulfills four functions of metabolism: (1) ATP generation, (2) catabolism, (3) anabolism, and (4) signaling.

enzymes, along with availability of energy, influence whether a metabolic reaction will proceed:

$$\text{Reactants} \xrightarrow{\text{Enzymes}} \text{Products} \, (\Delta \, \text{energy}).$$

It is important to realize that enzymes cannot determine the directionality of an individual reaction. Enzymes are proteins that serve as catalysts, which increase the rate of a reaction; they remain unaffected by the reaction itself. An enzyme will bind to a substrate and undergo a catalytic cycle that releases products. The enzyme itself remains intact and is poised to accept another substrate.

So, what determines whether a reaction requires or releases energy? Thermodynamics! I realize some readers are likely to shriek at the thought of thermodynamics. However, an appreciation of at least the basic thermodynamic principles is essential to understanding the directionality of metabolic pathways.

## BASICS OF THERMODYNAMICS

Thermodynamics is the study of energy, which is defined as the ability to do work. Potential and kinetic energy are two major forms of energy. In simple terms, kinetic energy is the energy caused by motion, whereas potential energy is stored energy. The law of conservation of energy states that energy can neither be created nor destroyed. The total energy in a system remains constant as dictated by the first law of thermodynamics. However, energy can change from one form to another. For example, cellular respiration converts the energy stored within glucose ($C_6H_{12}O_6$) to ATP in the presence of oxygen ($O_2$) and to carbon dioxide ($CO_2$), and water ($H_2O$) as products of the reaction. ATP is the major form of energy that living organisms use to conduct reactions. The potential energy of ATP can be converted to kinetic energy (e.g., to contract muscles). In biology, an example of potential energy is bond energy, which is the amount of energy required to break a chemical bond.

In the 1870s, the American physicist Josiah Willard Gibbs integrated into an equation the concepts of thermodynamics to determine whether a reaction requires energy or releases it:

$$\Delta G = \Delta H - T\Delta S.$$

At constant temperature and pressure, the Gibbs free-energy change, $\Delta G$, has two components known as the enthalpy change, $\Delta H$, and the entropy change, $\Delta S$. Enthalpy is a measure of heat content and entropy is measure of randomness. In biological reactions, the enthalpy of reactants and products is equal to their respective bond energies, and the overall change in enthalpy, $\Delta H$, is equal to the difference in bond energy between products and reactants. The current unit of measurement for enthalpy is the joule (J) in the International System of Units. However, most readers are likely familiar with the calorie as a unit of energy as it pertains to metabolism. A calorie is defined as the amount of heat energy required to increase the temperature of 1 g of water by 1°C at 1 atm of pressure. A calorie is ~4.184 joules. Entropy, $S$, is a measure of the degree of randomness of a system. According to the second law of thermodynamics, all processes in the universe incline toward randomness without input of energy. Thus, the only way to prevent the natural course of disorder or randomness is to put energy into a system. For example, living organisms are made of highly ordered structures and thus tend to have low disorder or entropy. This is primarily a result of the consumption of nutrients that help generate energy to counteract the disorder and therefore sustain life. However, once an organism ceases to take in nutrients and energy generation (ATP) is diminished, disorder (entropy) increases, resulting in the eventual death of the organism. The entropy change, $\Delta S$, of reactions is the change in randomness between products and reactants. At the cellular level, in catabolic reactions in which highly ordered structures, such as starch, are broken down, there is an increase in entropy ($\Delta S$ is positive), whereas in anabolic reactions, in which highly ordered structures, such as cholesterol, are generated, there is a decrease in entropy ($\Delta S$ is negative). Anabolic reactions require energy to counteract entropy for the generation of higher-order structures.

## WHAT DETERMINES WHETHER A REACTION REQUIRES ENERGY OR RELEASES ENERGY?

In an initial solution of reactants and products, a reaction that proceeds to produce more products than reactants will have $\Delta G < 0$, a reaction in which energy is released.

This reaction is referred to as exergonic or exoergic. The bond energy of reactants is greater than the products and thus $\Delta H < 0$. The products tend to be more disordered than the reactants; as a result, the reaction proceeds to a higher randomness state ($\Delta S > 0$). Plugging a negative $\Delta H$ and positive $\Delta S$ at constant temperature, T, into the Gibbs free-energy equation yields $\Delta G < 0$. The $\Delta G$ can be represented as the Gibbs free energy difference between products and reactants:

$$\text{exergonic reaction (energy released): } \Delta G = G_{\text{products}} - G_{\text{reactants}} < 0.$$

In an initial solution of reactants and products, a reaction that proceeds to produce more reactants than products will require external energy if the reaction is to accumulate products instead of reactants. This reaction is referred to as endergonic or endoergic. The bond energy of reactants is smaller than the products and thus $\Delta H > 0$. The products tend to be more ordered than the reactants, and thus the reaction proceeds to a higher-ordered state ($\Delta S < 0$). Plugging a positive $\Delta H$ and negative $\Delta S$ at constant temperature, T, into the Gibbs free-energy equation yields $\Delta G > 0$:

$$\text{endergonic reaction (energy required): } \Delta G = G_{\text{products}} - G_{\text{reactants}} > 0.$$

When the rates of forward and backward reactions are equal and the concentrations of products and reactions do not change, then $\Delta G = 0$; that is, the reaction is at equilibrium. Gibbs free energy can be represented graphically (Fig. 2-2). Gibbs free energy of biological reactions is expressed as kJ/mol.

It is vital to realize that Gibbs free energy does not account for the time it takes for a reaction to occur; rather, it determines whether the reaction releases or requires

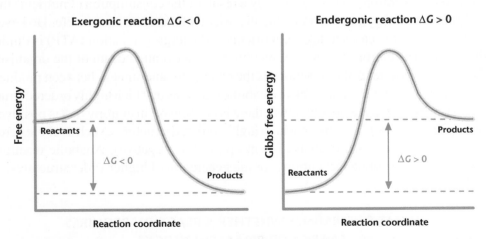

**Figure 2-2.** Exergonic and endergonic reactions. The change in Gibbs free energy of a reaction is the energy difference between products and reactants. Exergonic reactions have a negative change in the Gibbs free energy and, thus, release energy. In contrast, endergonic reactions have a positive change in the Gibbs free energy and, thus, require energy.

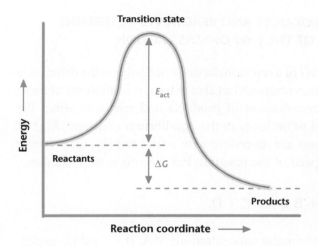

**Figure 2-3.** Transition state. The transition state is the highest point of energy in which an activated complex can either go back to the reactants or forward to the products. Activation energy ($E_{act}$) is the energy difference between reactants and the transition state.

energy. The rate of the reaction is governed by activation energy ($E_{act}$), expressed as kJ/mole, which is the energy required to push the reactants up to an activated complex known as the transition state. Hence, the activation energy is the energy difference between reactants and the activated complex. The peak of the energy profile is the transition state in which an activated complex can either go back to the reactants or go forward to the products (Fig. 2-3). The activation energy determines the rate of the reaction. Reactions with small activation energy proceed faster than reactions with high activation energy. Think of rolling a ball up a hill to the top. A smaller hill will require less time to roll the ball up than a larger hill. As we will learn later, enzymes as catalysts increase the rate of the reaction by lowering the activation energy (Fig. 2-4). Enzymes do not change the amount of Gibbs free energy of a reaction, but they do increase the rate of the reaction.

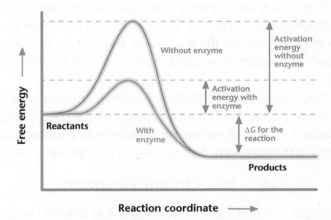

**Figure 2-4.** Enzymes lower the activation energy to increase the rate of reaction but not the overall change in Gibbs free energy of the reaction.

## CONCENTRATIONS OF PRODUCTS AND REACTANTS DETERMINE $\Delta G$ AS A RESULT OF THE LAW OF MASS ACTION

The law of mass action states that $\Delta G$ of a reaction depends not only on the difference in energy stored in the products versus reactants but also in the concentrations of these molecules. To appreciate how concentration of products and reactants affect the favorability of a reaction, we need to understand the equilibrium constant ($K_{eq}$) of a reaction. Most biological reactions are reversible with a forward ($k_1$) and reverse ($k_{-1}$) rate constant, which is the speed of the reaction. For the single-step reaction,

$$A + B \underset{k_{-1}}{\overset{k_1}{\rightleftharpoons}} C + D,$$

in which [A], [B], [C], and [D] are the molar concentrations of A, B, C, and D, respectively. The forward rate of a reaction is proportional to the product of the concentrations of A and B and is expressed as $k_1[A][B]$. The reverse rate of the reaction is proportional to the concentrations of C and D and is expressed as $k_{-1}[C][D]$.

At equilibrium, the forward rate of a reaction is equal to the reverse rate of the reaction:

$$k_1[A][B] = k_{-1}[C][D].$$

The equilibrium constant

$$K_{eq} = k_1/k_{-1} = [C][D]/[A][B].$$

The equilibrium constant is often used in Le Chatelier's principle (1884), which states that "Every change of one of the factors of an equilibrium occasions a rearrangement of the system in such a direction that the factor in question experiences a change in a sense opposite to the original change." In biological reactions, if the concentrations of products and reactants at equilibrium are changed, the equilibrium shifts in a direction that tends to restore the original concentrations (to counteract the change). For example, if you add the reactants A and B in the reaction, then the reaction goes forward. In contrast, if you add products C and D, then the reaction goes backward.

The actual free energy, $\Delta G$ (kJ/mol), of a given reaction is the standard free-energy change referred to as $\Delta G^{o\prime}$, which depends on the bond energy and disorder of the products and reactants, along with the concentration of the products and reactants. Gibbs derived the following equation to express this relationship:

$$\Delta G = \Delta G^{o\prime} + RT \ln [C][D]/[A][B],$$

in which R is the gas constant 8.315 J/mol·K and T is absolute temperature.

The $\Delta G^{o\prime}$ can be calculated by setting up a reaction under standard physiological conditions and then allowing it to proceed to equilibrium in which the concentrations of all reactants and products are measured. The change in actual free energy is $0$ ($\Delta G = 0$), and $\Delta G^{\circ\prime}$ can be calculated directly from $K_{eq}$:

$$0 = \Delta G^{\circ\prime} + RT \ln [C][D]/[A][B],$$
$$\Delta G^{\circ\prime} = -RT \ln [C][D]/[A][B].$$

## HOW DO ENDERGONIC REACTIONS UNDER EQUILIBRIUM CONDITIONS BECOME EXERGONIC REACTIONS?

First, an endergonic reaction under equilibrium conditions ($\Delta G^{\circ\prime} > 0$) in a cell can become exergonic if the concentration of products is very small compared with the concentration of reactants. In this scenario, [products]/[reactants], referred to as the mass-action ratio, will be $<1$. The natural log (ln) of a number that is $<1$ is a negative number and will result in the actual Gibbs free energy, $\Delta G$, to be negative. Second, an endergonic reaction is coupled to an exergonic reaction. The conversion of ATP to ADP and $P_i$ is a favorable reaction and is often used to drive unfavorable reaction. Let us look at the example of ATP $\rightarrow$ ADP + $P_i$ to see how concentration of ATP/ADP can make the difference in the amount of Gibbs free energy available to drive unfavorable reactions.

An example of an ATP-coupled reaction is (also see Fig. 2-5)

$$\Delta G^{\circ\prime} \text{ (kJ/mol)}$$

Glucose + $P_i$ → Glucose-6- phosphate + $H_2O$ + 13.8

ATP + $H_2O$ → ADP + $P_i$ − 30.5

_ _ _ _ _ _ _ _ _ _ _ _ _ _ _ _ _ _ _ _ _ _ _ _ _ _

ATP + Glucose (Glu) → Glucose-6-Phosphate + ADP − 16.7.

This first reaction of glycolysis (see Chapter 3) is an irreversible reaction that commits glucose to this metabolic pathway. The conversion of glucose to glucose

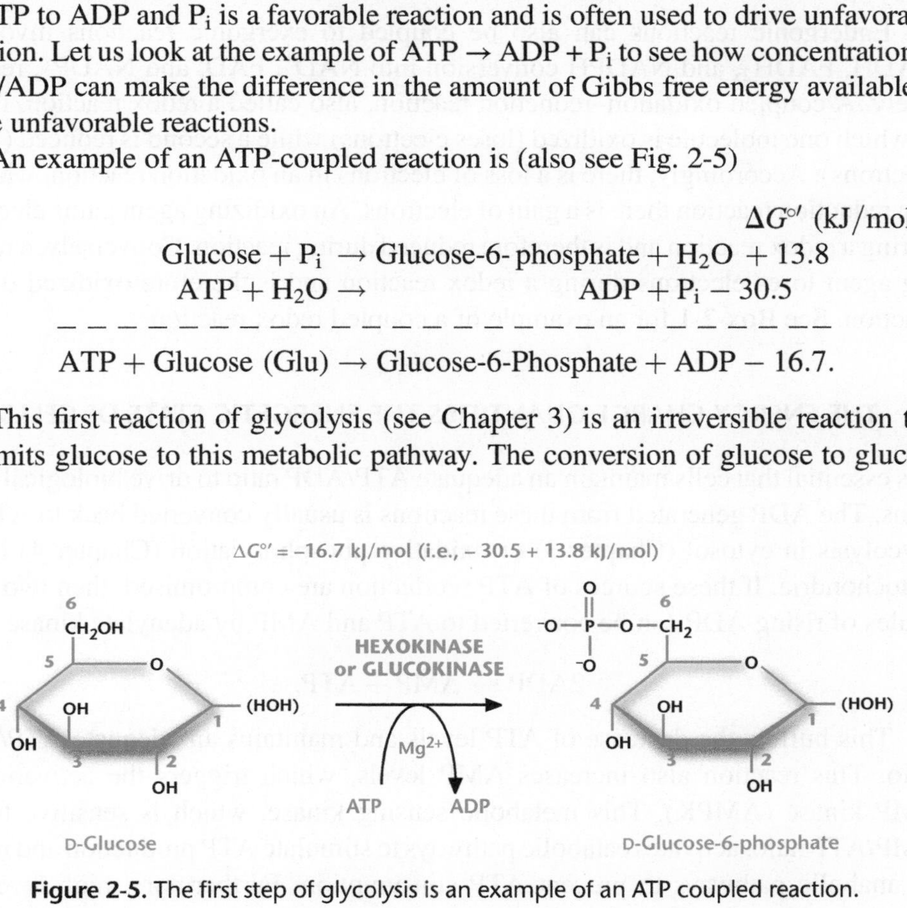

$$\Delta G^{\circ\prime} = -16.7 \text{ kJ/mol (i.e., } -30.5 + 13.8 \text{ kJ/mol)}$$

Figure 2-5. The first step of glycolysis is an example of an ATP coupled reaction.

6-phosphate is a reaction and has a $\Delta G^{\circ\prime} = +13.8$. However, coupling this reaction to the conversion of ATP to ADP and $P_i$ results in an exergonic reaction with a $\Delta G^{\circ\prime} = -30.5$. Thus, under standard conditions, the $\Delta G^{\circ\prime}$ is $-16.7$, but remember that the overall $\Delta G$ for the reaction will be dependent on the ratio of products/reactants along with the $\Delta G^{\circ\prime}$. The overall $\Delta G$ is even more favorable because the reactants, glucose and ATP, are much more in excess than the products, glucose 6-phosphate and ADP. By knowing the $\Delta G^{\circ\prime}$ and metabolite concentration, we plug in these values to $\Delta G = \Delta G^{\circ\prime} + RT \ln (\text{products/reactants})$. If we use a steady-state concentration of products (glucose 6-phosphate and ADP) and reactants (Glu and ATP) observed in red blood cells, then the equation looks like

$$\Delta G = \Delta G^{\circ\prime} + RT \ln \frac{([\text{Glucose 6-phosphate}][\text{ADP}][P_i])}{([\text{Glu}][\text{ATP}])},$$

$$\Delta G = -16.7 \text{ kJ/mol} + (8.314 \text{ J/mol K})(310 \text{ K}) \ln \frac{([0.083][0.14][1])}{([5][1.85])},$$

$$\Delta G = -33 \text{ kJ/mol}.$$

Endergonic reactions can also be coupled to exergonic reactions involving NADH, $FADH_2$, and NADPH conversion into $NAD^+$, FAD, and $NADP^+$, respectively. A coupled oxidation–reduction reaction, also called a redox reaction, is one in which one molecule is oxidized (loses electrons) while a second is reduced (gains electrons). Accordingly, there is a loss of electrons in an oxidation reaction, whereas in a reduction reaction there is a gain of electrons. An oxidizing agent gains electrons during a redox reaction and is therefore reduced during reaction. Conversely, a reducing agent loses electrons during a redox reaction and is therefore oxidized during reaction. See Box 2-1 for an example of a coupled redox reaction.

## THE ENERGY CHARGE QUANTIFIES THE ENERGETIC STATE OF CELLS

It is essential that cells maintain an adequate ATP/ADP ratio to drive biological reactions. The ADP generated from these reactions is usually converted back to ATP by glycolysis in cytosol (Chapter 3) or oxidative phosphorylation (Chapter 4) in the mitochondria. If these sources of ATP production are compromised, then two molecules of rising ADP can be converted to ATP and AMP by adenylate kinase:

$$2\text{ADP} \leftrightarrow \text{AMP} + \text{ATP}.$$

This buffers the decrease of ATP levels and maintains an adequate ATP/ADP ratio. This reaction also increases AMP levels, which triggers the activation of AMP kinase (AMPK). This metabolic sensing kinase, which is sensitive to the AMP/ATP ratio, activates catabolic pathways to stimulate ATP production and inhibits anabolic pathways to prevent ATP consumption. Furthermore, high levels of

## BOX 2-1. A COUPLED REDOX REACTION

An example of a coupled redox reaction is the conversion of pyruvate to lactate. The conversion of pyruvate to lactate is enzymatically catalyzed by lactate dehydrogenase. In this reaction, pyruvate gains two electrons and is reduced to lactate; that is, in the reduction half of the reaction, NADH loses two electrons and is oxidized to $NAD^+$ (i.e., in the oxidizing half of the reaction).

This redox reaction consists of two half-reactions

$$\textbf{pyruvate} + \textbf{2H}^+ + \textbf{2e}^- \rightarrow \textbf{lactate} \qquad \text{(pyruvate gains e}^-\text{[reduction half reaction])},$$
$$\textbf{NADH} + \textbf{H}^+ \rightarrow \textbf{NAD}^+ + \textbf{2H}^+ + \textbf{2e}^- \quad \text{(NADH loses e}^-\text{[oxidation half reaction])},$$
$$\textbf{pyruvate} + \textbf{NADH} + \textbf{H}^+ \rightarrow \textbf{lactate} + \textbf{NAD}^+ \quad \text{(net reaction).}$$

NADH donates (or loses) two electrons and is oxidized to $NAD^+$. NADH is the reducing agent. Pyruvate accepts (or gains) two electrons and is reduced. Pyruvate is the oxidizing agent. These half reactions, measured in volts, are the reduction potential, $E^{o\prime}$, of a redox reaction. Reduction potential is a measure of the propensity of a metabolite to gain electrons. Thus, the $E^{o\prime}$ of the two reduction half-reactions of pyruvate conversion to lactate are as follows:

(1) $\quad$ pyruvate $+ 2H^+ + 2e^- \rightarrow$ lactate, $\quad E^{o\prime} = -0.190$ V,

(2) $\quad NAD^+ + 2H^+ + 2e^- \rightarrow$ NADH $+ H^+$, $\quad E^{o\prime} = -0.320$ V.

The higher (less negative) the $E^{o\prime}$, the more likely it has a propensity to gain electrons. The pyruvate/lactate pair has a higher $E^{o\prime}$ than the $NAD^+$/NADH pair and is likely to accept more electrons. Thus, the $NAD^+$ reaction is reversed along with the sign of the reduction potential:

(2 reversed) $\quad$ NADH $+ H^+ \rightarrow NAD^+ + 2H^+ + 2e^-$, $\quad E^{o\prime} = 0.320$ V.

Then, reaction (1) and the reversed reaction (2) can be added together to arrive at the net redox reaction, and the $\Delta E^{o\prime}$ can be found by adding the reduction potential of the first half-reaction to the oxidation potential of the second half-reaction

| | |
|---|---|
| (1) pyruvate $+ 2H^+ + 2e^- \rightarrow$ lactate, | $E^{o\prime} = -0.190$ V, |
| (2 reversed) NADH $+ H^+ \rightarrow NAD^+ + 2H^+ + 2e^-$, | $E^{o\prime} = 0.320$ V, |
| (sum) pyruvate $+$ NADH $+ H^+ \rightarrow$ lactate $+ NAD^+$, | $\Delta E^{o\prime} = 0.130$ V (sum). |

This $\Delta E^{o\prime}$ of the redox reaction can be converted into $\Delta G^{o\prime}$, in which $n$ is the number of electrons involved in the reaction and F is the Faraday constant (96.5 kJ/volt∗mol):

$$\Delta G^{o\prime} = -nF\Delta E^{o\prime}.$$

If we plug $\Delta E^{o\prime} = 0.130$ V into the above equation

$$\Delta G^{o\prime} = -2(96.5 \text{ KJ/V mol})(0.130 \text{ V}),$$

we get $\Delta G^{o\prime} = -25$ kJ/mol. Because $\Delta G^{o\prime} < 0$, this is a reaction that releases energy. Note that these redox reactions have a positive $\Delta E^{o\prime}$ and are reflected as negative $\Delta G^{o\prime}$. The equation above reflects standard conditions. Remember from the previous section that the actual $\Delta G$ in a cell will be determined by the ratio of the reactants (pyruvate and NADH) to products

*(Continued)*

---

**BOX 2-1.** *(Continued)*

---

(lactate and NAD$^+$). To calculate *E* of redox reactions taking into the concentrations of reactants and products, we use the Nernst equation:

$$\Delta E = \Delta E^{\circ\prime} - (RT/nF) \ln ([\text{products}]/[\text{reactants}]).$$

Subsequently, $\Delta G$ can be calculated by $-nF\Delta E$.

---

AMP or ATP activate or inhibit, respectively, metabolic reactions involved in generating ATP.

In the 1960s, Daniel Atkinson developed a formula to quantitatively estimate the energy state of the cell based on the concentrations of ATP, ADP, and AMP, referred to as the energy charge:

$$\text{energy charge} = \frac{[\text{ATP}] + \frac{1}{2}[\text{ADP}]}{[\text{ATP}] + [\text{ADP}] + [\text{AMP}]}.$$

An energy charge ranges from 0 (all AMP) to 1 (all ATP). If ATP, ADP, and AMP have the same concentration, then the energy charge is 0.5. Cells maintain an energy charge between 0.7 and 0.9 by regulating metabolic flux, which is the turnover rate of reactants and products through pathways that generate and consume ATP. An imbalance between rates of use and regeneration of ATP causes a decrease in the energy charge. Consequently, a lower energy charge value slows down macromolecular synthesis and growth. Atkinson's important insight was that cells counteract a decreasing energy charge by diminishing the rate of ATP-consuming reactions and increasing the rate of ATP-regenerating reactions.

## ENZYMES CATALYZE METABOLIC PATHWAYS

Enzymes can be either a protein or a catalytic RNA, such as a ribozyme (RNA enzyme). Enzymes increase the rate of reactions by lowering the activation energy of reactions. The rate of reaction ($k$) is how fast a reactant gets used up or how fast a product gets produced. Without enzymes, the rate of metabolic reactions would proceed too slowly to deliver the necessary energy, as well as the biosynthesis of macromolecules to keep organisms alive. An important consideration to remember is that if a reaction is not thermodynamically favorable, then enzymes cannot make the reaction go forward. Enzymes do not alter the Gibbs free energy of a reaction (Fig. 2-4). Imagine pushing a heavy ball up a hill. It will take less time to push the ball over a smaller hill than a larger hill. However, the strength to push the ball over either hill is similar. If the person does not have the strength to push the ball, it does not matter if the hill is big or small.

Many enzymes also require the presence of nonprotein molecules called cofactors for their catalytic activity. An inactive enzyme without the cofactor is designated as an apoenzyme, whereas an active enzyme bound with a cofactor is referred to as a holoenzyme. Cofactors, such as heme, that are permanently bound to enzymes are called prosthetic groups, and those, such as $NAD^+$, that are not permanently bound are termed coenzymes. Many enzymes also use metal ions, such as iron, copper, or magnesium, for their catalytic activity. Enzymes show specificity to the reactions they catalyze. They usually have specific substrates (reactants); however, they are not consumed like a substrate of a reaction nor do they appear as reaction products. A substrate and an enzyme form a transient intermediate before the generation of the product. Once the product is generated, the enzyme returns to its original form poised to accept another substrate.

A basic enzymatic reaction is

$$S \quad + \quad E \quad \longrightarrow \quad ES \quad \longrightarrow \quad P \quad + \quad E.$$

| Substrate | Enzyme | Enzyme–Substrate | Product | Enzyme |

However, most biological reactions are reversible; thus, the enzyme–substrate complex can either become a product or substrate, depending on their respective concentrations. Looking at this reaction, it is apparent that both the substrate and enzyme concentrations are going to affect the rate of the reaction. The other important factors that affect the rate of reactions are temperature, pH, and the presence of activators or inhibitors. Enzymes have a temperature and pH range at which their activity is greatest.

There are two important points regarding enzyme kinetics at a constant pH and temperature.

1. If there is unlimited substrate, then the rate of a reaction is proportional to enzyme concentration (Fig. 2-6).

2. If the enzyme concentration is constant, then the initial rate of the reaction ($V_0$) will increase as the substrate concentration increases from 0 until it reaches a maximum, defined as $V_{max}$. After this point, a further increase in the substrate concentration will not increase the rate. The $K_m$ of the reaction is defined as the substrate concentration at $\frac{1}{2}V_{max}$ (Fig. 2-6). Quantitatively, this is illustrated in the Michaelis–Menten equation:

$$v_0 = \frac{V_{max}[S]}{K_m + [S]}.$$

Plugging in a substrate concentration $[S] = K_m$ in the equation gives $v_0 = V_{max}/2$. The $K_m$ is also referred to as the Michaelis–Menten constant.

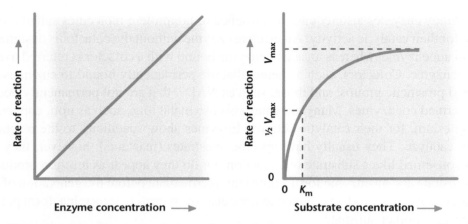

**Figure 2-6.** Rate of a reaction is dependent on the enzyme and substrate concentration.

The size of $K_m$ tells us the following things about a particular enzyme.

* A small $K_m$ indicates that the enzyme requires only a small amount of substrate to become saturated. Hence, the maximum velocity is reached at relatively low substrate concentrations.

* A large $K_m$ indicates the need for high substrate concentrations to achieve maximum reaction velocity.

## REGULATION OF ENZYMES AFFECTS THEIR CATALYTIC ACTIVITY

Aside from increasing rate of reaction and having specific substrates, enzymes are regulated so that they can go from a low- to a high-activity state. Enzymes have an active site in which the substrate binds to form the enzyme–substrate complex and an allosteric site at which metabolites, ions, and proteins can bind and affect the catalytic activity of the enzyme by increasing or decreasing the substrate–enzyme interaction. Allosteric enzymes, a special class of enzymes, oscillate between two conformations: a high-activity, high-affinity relaxed (R) state and a low-activity, low-affinity tense (T) state. These enzymes are made of multiple subunits and require an activator to bind all the subunits to stabilize the active form. There are also multiple mechanisms to inhibit enzyme activity, including feedback, competitive, and noncompetitive inhibitions (Fig. 2-7). Feedback inhibition is a mechanism by which product accumulation inhibits the enzyme. In competitive inhibition, an inhibitor mimics the substrate and competes for binding to the active site. This forces the natural substrate for the enzyme to increase its concentration to achieve $V_{max}$. However, the presence of the inhibitor does not change the $V_{max}$ of the reaction because excess natural substrate would compete away the inhibitor. Thus, competitive inhibition results in an increase in $K_m$ without changing $V_{max}$. In contrast, noncompetitive

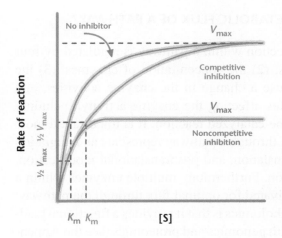

**Figure 2-7.** Competitive versus noncompetitive inhibition of enzymes. Competitive inhibition results in an increase in $K_m$ without changing $V_{max}$, whereas noncompetitive inhibition causes a decrease in $V_{max}$ without changing $K_m$.

inhibition involves an inhibitor that binds to an allosteric site to deactivate the enzyme. The natural substrate can still bind to the enzyme, but as long as the inhibitor is bound to the allosteric site, the enzyme is essentially catalytically inactive. Thus, the noncompetitive inhibitor decreases $V_{max}$ and addition of excess natural substrate cannot overcome inhibition.

## ILLUSTRATION OF CONTROL OF FLUX THROUGH A HYPOTHETICAL METABOLIC PATHWAY

Let us take a look at a hypothetical simple metabolic pathway to integrate the concept of how enzymes and metabolites control flux, defined as rate of turnover of molecules through a metabolic pathway. A, B, C, and D are defined as metabolites and E1, E2, and E3 are their enzymes. The first step catalyzed by enzyme E1 is irreversible and commits the initial reactant A to this pathway. The irreversible steps catalyzed by E1 and E3 also become the regulatory steps in the pathway. The middle reaction, carried by E2, is reversible. Because C is quickly removed by E3, this forces B to become C. Alternatively, if E3 becomes inhibited at an allosteric site, then C would build up and be converted back to B because the E2-catalyzed reaction is reversible. This could prevent any further conversion of A to B by E1 because B could inhibit E1 (i.e., product inhibition):

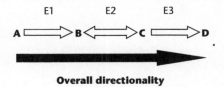

**Overall directionality**

In summary, always look at the concentration of the products and reactants, along with the enzyme activity, to get a sense of the direction of any given reaction.

## WHAT DETERMINES THE METABOLIC FLUX OF A PATHWAY?

Metabolic flux through a particular direction within a pathway is regulated by four factors: (1) the availability of reactants, (2) the concentration of enzymes, (3) the posttranslational modifications that cause a change in the enzyme activities, and (4) the concentrations of the metabolites affecting the enzyme activity, including the reactants and products of the enzyme-catalyzed reaction. It is important to note enzyme function, and so metabolic flux through a pathway represents an integration of enzyme gene expression, protein translation, and posttranslational modification, as well as enzyme–metabolite interaction. Furthermore, multiple enzymes within a metabolic pathway are coordinately activated for optimal flux through the pathway. Therefore, one advantage of doing metabolomics is that it provides a functional read-out of metabolic pathways compared with genomics and proteomics (see the Appendix). In Chapters 3–9, the major regulation of enzymes discussed is restricted to how metabolites, such as adenine nucleotides, regulate enzyme function. Chapters 10 and 11 cover the signaling pathways and transcription factors that regulate gene expression, protein translation, and posttranslational modification of enzymes to control anabolic and catabolic processes.

## ADDITIONAL READING

Alberty RA. 2003. *Thermodynamics of biochemical reactions*. Wiley-Interscience, Hoboken, NJ.

Atkinson DE. 1968. The energy charge of the adenylate pool as a regulatory parameter. Interaction with feedback modifiers. *Biochemistry* **7:** 4030–4034.

Gibbs JW. 1875–1878. On the equilibrium of heterogeneous substances. Transactions of the Connecticut Academy 1876. In *The collected works of J. Willard Gibbs*. Yale University Press, New Haven, CT.

Johnson KA, Goody RS. 2011. The original Michaelis constant: Translation of the 1913 Michaelis–Menten paper. *Biochemistry* **50:** 8264–8269.

Krebs HA, Kornberg HL. 1957. *Energy transformations in living matter*. Springer, Berlin.

Nielsen J. 2003. It is all about metabolic fluxes. *J Bacteriol* **185:** 7031–7035.

NIST Standard Reference Database 74. Biochemical Science Division, National Institute of Standards and Technology, Gaithersburg, MD. http://xpdb.nist.gov/enzyme_thermodynamics (for thermodynamic data on enzyme-catalyzed reactions).

Speight J. 2004. *Lange's handbook of chemistry*, 70th anniversary edition, 16th ed. McGraw-Hill, New York (for tables of values).

Walsh C. 1979. *Enzymatic reaction mechanisms*. WH Freeman, San Francisco.

# 3

## Glycolysis

GLYCOLYSIS IS AN ANCIENT PATHWAY THAT EVOLVED WELL BEFORE oxygen was present in the Earth's atmosphere and is highly conserved among living organisms. Glycolysis was the first metabolic pathway elucidated and is also referred to as the Embden–Meyerhof–Parnas pathway (see Box 3-1). The word "glycolysis" is derived from the Greek "glykys," meaning "sweet," and "lysis," which means "to split." This refers to the splitting of one glucose molecule into two molecules of pyruvate, the end product of glycolysis. In the presence of oxygen, pyruvate usually enters the mitochondria where it is oxidized to acetyl-CoA, whereas in the absence of oxygen, pyruvate is reduced into lactate. Glycolysis involves 10 reactions that take place in the cytosol and generates two ATP molecules without the requirement of molecular oxygen. In contrast, oxidative phosphorylation in the mitochondria generates 30 ATP molecules but requires oxygen (see Chapter 4). Multiple simple sugars can enter glycolysis, including the monosaccharides glucose, galactose, and fructose. Most of us ingest these simple sugars through consumption of products that contain the disaccharides sucrose (table sugar) or lactose (milk sugar). Sucrose is composed of one molecule of glucose and fructose, whereas lactose contains one molecule of glucose and galactose. The digestive enzymes sucrase and lactase break down sucrose and lactose into simple sugars.

The overall reaction of glycolysis is exergonic ($\Delta G = -96$ kJ/mol in erythrocytes):

$$\text{glucose} + 2NAD^+ + 2ADP + 2P_i \rightarrow 2\text{pyruvate} + 2NADH + 2H^+ + 2ATP + 2H_2O.$$

### THERE ARE THREE MAJOR FEATURES OF GLYCOLYSIS

1. It is the only pathway that can generate ATP in the absence of oxygen (anaerobic conditions) or in cells lacking mitochondria, such as red blood cells (erthyrocytes). In these scenarios, pyruvate is converted into lactate.

2. In the presence of oxygen, glycolysis generates pyruvate, which enters the TCA cycle (also called the citric acid cycle and the Krebs cycle) located within mitochondria to produce ATP.

---

**BOX 3-1. OTTO FRITZ MEYERHOF AND ENERGY TRANSFORMATION IN THE CELL**

---

Archibald Vivian Hill described Otto Fritz Meyerhof (1884–1951) as "...always been betwixt and between: a physiological chemist or a chemical physiologist, perhaps we should call him a 'chemiologist'." These characteristics are precisely what enabled Meyerhof and his co-workers to dissect the glycolysis pathway. Meyerhof had won the 1922 Nobel Prize in Physiology or Medicine (awarded in 1923 because of a quirk of Alfred Nobel's will) "for his discovery of the fixed relationship between the consumption of oxygen and the metabolism of lactic acid in the muscle" with Hill "for his discovery relating to the production of heat in the muscle." Meyerhof initially thought he would pursue psychiatry and philosophy, but in 1909 he crossed paths with Otto Warburg, who persuaded him to study physiology. Early in his career, Meyerhof believed that the laws of physics and chemistry should apply to living organisms, and in 1913 he lectured on a theory of the thermodynamics of living matter—"The Energetics of Cell Processes." In the late 1920s through the 1930s at the Institute of Physiology of the Kaiser Wilhelm Institute of Medical Research in Heidelberg, Meyerhof put together several key discoveries, including Gustav Embden's isolation of AMP and his outline of the glycolysis pathway (just before his death), Jakub Parnas's work on phosphorolysis, and Karl Lohman's discovery of ATP, and combined them with precise laboratory work to dissect and rebuild the glycolysis pathway and identify one-third of the enzymes involved. By identifying these intermediate reactions and showing the series of transformations that make energy available to the cell, Meyerhof answered the questions posed in his 1913 lecture about how energy transformations and chemical changes affect the function of cells. He later confirmed that the glycolysis pathway in muscle and yeast was the same, thus showing it to be an essential pathway in living organisms. That Meyerhof was much respected as a mentor and investigator is reflected in the "who's who" of researchers who passed through his laboratories in Germany—among them, David Nachmansohn, Severo Ochoa, Fritz Lipmann, George Wald, Andre Lwoff, Fritz Haber, and Otto Kahn.

3. Many of the metabolites of glycolysis and the TCA cycle can also enter anabolic pathways that generate NADPH and the building blocks needed for generation of glycogen, lipid, nucleotide, and protein synthesis. The biosynthetic pathways that glycolytic intermediates funnel into include the pentose phosphate, the hexosamine, and the serine and glycerol biosynthetic pathways.

## QUICK GUIDE TO THE ENERGY-GENERATING CAPACITY OF GLYCOLYSIS (FIG. 3-1)

- Glycolysis takes place in the cytosol and does not require oxygen to generate ATP. Note that there is no net loss of carbon or oxygen atoms in glycolysis.

- The 10 enzymatic reactions can be divided into two phases: ATP investment (reactions 1–5) and ATP payoff (reactions 6–10).

- Every one molecule of glucose entering glycolysis generates two molecules of glyceraldehyde 3-phosphate using two molecules of ATP during the ATP investment phase.

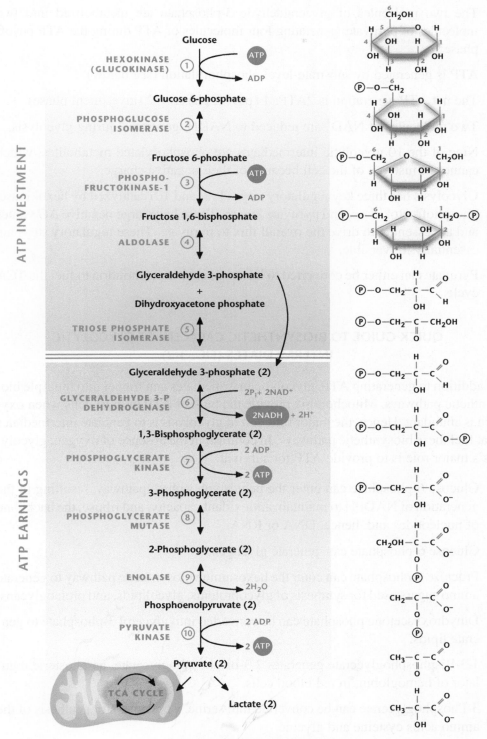

**Figure 3-1.** The 10 steps of glycolysis.

- The two molecules of glyceraldehyde 3-phosphate are metabolized into two molecules of pyruvate, generating four molecules of ATP during the ATP payoff phase.

- ATP is generated by substrate-level phosphorylation (see below).

- The net ATP generation is 2ATP: 4 (payoff phase) – 2 (investment phase).

- Two molecules of $NAD^+$ are reduced to NADH, generated during glycolysis.

- Nine of the 10 glycolytic intermediates are phosphorylated metabolites, which cannot diffuse out of the cell because of their negative charge.

- Glycolysis has three key regulatory steps (1, 3, and 10) catalyzed by hexokinase, phosphofructokinase, and pyruvate kinase. These have large negative $\Delta G$ values and are essential to drive the overall flux to pyruvate. These regulatory steps are essentially irreversible.

- Pyruvate can either be converted to lactate or enter mitochondria to fuel the TCA cycle.

## QUICK GUIDE TO BIOSYNTHETIC CAPACITY OF GLYCOLYTIC INTERMEDIATES (FIG. 3-2)

In addition to generating ATP, glycolytic intermediates can funnel into multiple biosynthetic pathways. Mitochondria provide the bulk of ATP in most cells when oxygen is abundant. Hence, the major function of glycolysis is to generate intermediates that fuel these biosynthetic pathways. In contrast, in the absence of oxygen, glycolysis's major role is to provide ATP for survival.

- Glucose 6-phosphate can enter the pentose phosphate pathway, resulting in the generation of NADPH to maintain antioxidant capacity, and ribose, the backbone of nucleotides and, hence, DNA or RNA.

- Glucose 6-phosphate can generate glycogen.

- Fructose 6-phosphate can enter the hexosamine biosynthetic pathway to generate amino sugars used for synthesis of glycoproteins, glycolipids, and proteoglycans.

- Dihydroxyacetone phosphate can be converted into glycerol 3-phosphate to generate lipids.

- 1,3-Bisphosphoglycerate generates 2,3-bisphosphoglycerate, an allosteric regulator of hemoglobin, in red blood cells.

- 3-Phosphoglycerate can be converted into serine, a precursor for synthesis of the amino acids cysteine and glycine.

- Pyruvate can generate the amino acid alanine.

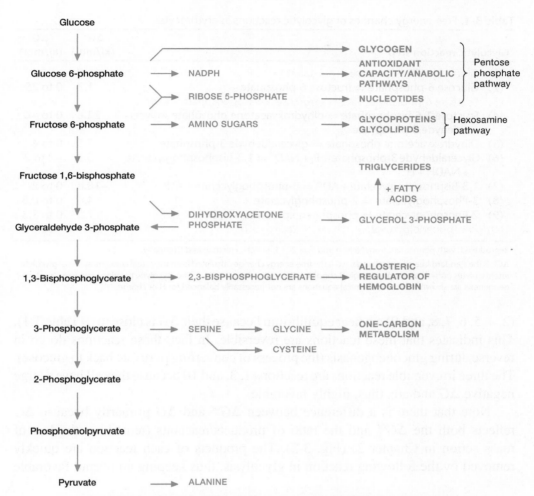

**Figure 3-2.** Glycolytic intermediates funnel into biosynthetic pathways.

## WHAT DRIVES THE 10 REACTIONS OF GLYCOLYSIS FORWARD?

Remember from our discussion of thermodynamics in Chapter 2 that the directionality of reactions is governed by initial reactant (glucose), enzyme concentration, and activity, as well as the $\Delta G$ defined as $\Delta G^{\circ\prime}$ + RT ln([products]/[reactants]). Glucose concentration in the blood is typically ~5 mM, which is at a higher concentration than the $K_m$ of basal glucose transporters and hexokinases (step 1 of glycolysis) in most cells. The Gibbs free energy, $\Delta G$, has to be either negative or close to 0 if glycolysis is to proceed in the forward direction. The $\Delta G^{\circ\prime}$ and $\Delta G$ of all 10 reactions are provided in Table 3-1. $\Delta G$ calculations are based on the steady-state metabolite concentration of products and reactants of each reaction in red blood cells and, thus, are different than the $\Delta G^{\circ\prime}$. A careful examination of the table indicates that most of the reactions

**Table 3-1.** Free-energy changes of glycolytic reactions in erythrocytes

| Glycolytic reaction step | $\Delta G^{o\prime}$ (kJ/mol) | $\Delta G$ (kJ/mol) |
|---|---|---|
| (1)  Glucose + ATP → glucose 6-phosphate + ADP | −16.7 | −33.4 |
| (2)  Glucose 6-phosphate ⇌ fructose 6-phosphate | 1.7 | 0 to 25 |
| (3)  Fructose 6-phosphate + ATP → fructose 1,6-bisphosphate + ADP | −14.2 | −22.2 |
| (4)  Fructose 1,6-bisphosphate ⇌ dihydroxyacetone phosphate + glyceraldehyde 3-phosphate | 23.8 | 0 to −6 |
| (5)  Dihydroxyacetone phosphate ⇌ glyceraldehyde 3-phosphate | 7.5 | 0 to 4 |
| (6)  Glyceraldehyde 3-phosphate + $P_i$ + $NAD^+$ ⇌ 1,3-bisphosphoglycerate + NADH + $H^+$ | 6.3 | −2 to 2 |
| (7)  1,3-Bisphosphoglycerate + ADP ⇌ 3-phosphoglycerate + ATP | −18.8 | 0 to 2 |
| (8)  3-Phosphoglycerate ⇌ 2-phosphoglycerate | 4.4 | 0 to 0.8 |
| (9)  2-Phosphoglycerate ⇌ phosphoenolpyruvate + $H_2O$ | 7.5 | 0 to 3.3 |
| (10)  Phosphoenolpyruvate + ADP → pyruvate + ATP | −31.4 | −16.7 |

Reproduced, with permission, from Nelson and Cox 2013, © W.H. Freeman and Company.

$\Delta G^{o\prime}$ is the standard free-energy change. $\Delta G$ is the free-energy change calculated from the actual concentrations of glycolytic intermediates present under physiological conditions in erythrocytes, at pH 7. The glycolytic reactions bypassed in gluconeogenesis are shown in red. Biochemical equations are not necessarily balanced for H or charge.

(2, 4, 5, 6, 7, 8, and 9) are near equilibrium because their $\Delta G$ is close to 0 (Table 3-1). This indicates that these reactions are reversible. In fact, these reactions do go in reverse during gluconeogenesis (the process of converting pyruvate back to glucose). The three irreversible reactions are reactions 1, 3, and 10 because they all have a large negative $\Delta G$ and are, thus, highly favorable.

Note that there is a difference between $\Delta G^{o\prime}$ and $\Delta G$ primarily because $\Delta G$ reflects both the $\Delta G^{o\prime}$ and the ratio of products/reactants (remember the law of mass action in Chapter 2) (Fig. 3-3). The products of each reaction are quickly removed by the following reaction in glycolysis, thus keeping an overall favorable

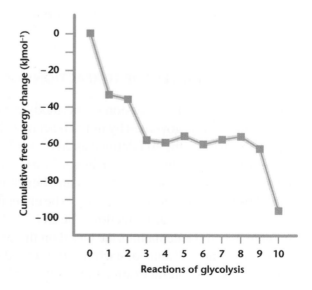

**Figure 3-3.** Gibbs free-energy change in 10 steps of glycolysis. The free-energy changes are based on Table 3-1. Note that the Gibbs free-energy changes in reactions 6–10 are multiplied by 2 because one molecule of glucose is converted into two molecules of glyceraldehyde 3-phosphate.

Gibbs free energy. For example, take a look at reaction 4, which requires a large positive $\Delta G^{o\prime}$ (+23.8) to drive the reaction forward, resulting in catalysis of a six-carbon fructose 1,6-bisphosphate into two three-carbon molecules, glyceraldehyde 3-phosphate and dihydroxyacetone phosphate. However, these three-carbon molecules are quickly removed by the subsequent reactions of glycolysis, ensuring a low ratio between the concentrations of product and reactants and a negative $\Delta G$. In fact, we look at the cumulative free-energy change from each reaction, and then we arrive at a favorable overall Gibbs free-energy change for glycolysis at –96 kJ/mol in erythrocytes (Fig. 3-3).

## WHY USE ATP TO MAKE ATP?

A puzzling aspect of glycolysis is the use of 2 ATP in the first phase (ATP investment) and the generation of 4 ATP in the second phase (ATP payoff). Why consume 2 ATP in the first place? There are two explanations. First, the phosphorylation of glucose using ATP makes the glycolytic intermediates into phosphorylated metabolites, which cannot diffuse out of the cell because of their negative charge. Second, the phosphorylated metabolites can be used for substrate-level phosphorylation, a process in which a phosphoryl ($PO_3$) group is transferred from a phosphorylated metabolite, like 1,3-bisphosphoglycerate, to ADP to generate ATP. The generation of ATP requires +30.5$\Delta G^{o\prime}$ when bound to $Mg^{2+}$ and has to be coupled to reactions that have a large negative $\Delta G^{o\prime}$. There are two glycolytic reactions that have a large negative $\Delta G^{o\prime}$ and are coupled to the generation of ATP. In step 7, the conversion of 2,3-bisphoshoglycerate to 2-phosphoglycerate has a $\Delta G^{o\prime}$ of –37.6 kJ/mol, and in step 10 the conversion of 2 phosphoenolpyruvate to 2 pyruvate has a $\Delta G^{o\prime}$ of –62.8 kJ/mol. To generate these three-carbon phosphate molecules from glucose, the first phase of glycolysis phosphorylates glucose using two ATP molecules and converts it from one six-carbon glucose molecule into two three-carbon glyceraldehyde 3-phosphate molecules. Subsequently, glyceraldehyde 3-phosphate dehydrogenase couples $NAD^+$ reduction to NADH to the addition of inorganic phosphate to glyceraldehyde-3-phosphate, resulting in the generation of 1,3-bisphosphoglycerate, whereas subsequent reactions produce phosphoenolpyruvate.

In the presence of oxygen, NADH and pyruvate are transported into the mitochondria where ATP is generated, as discussed in the next section. It is important to note that under ample oxygen conditions, glycolysis serves to provide intermediates that can fuel the biosynthetic pathways in most cells. Glycolytic ATP generation is paramount when oxygen is limiting to the generation of mitochondrial-dependent ATP. There are exceptions, such as erythrocytes that contain no mitochondria or cells such as neutrophils and endothelial cells that contain few mitochondria. In these cases, the flux through glycolysis is high enough to sustain both ATP generation and fueling of the biosynthetic pathways.

### WHAT HAPPENS TO THE NADH AND PYRUVATE GENERATED DURING GLYCOLYSIS?

The fate of NADH and pyruvate is intimately linked. In the presence of ample oxygen, both pyruvate and NADH are transported into the mitochondria, where pyruvate enters the TCA cycle while NADH is oxidized to $NAD^+$ by the electron transport chain (see Chapter 4) and transported back into the cytosol to promote glycolysis (Fig. 3-4). The transport of NADH and $NAD^+$ in and out of mitochondria involves intricate shuttling mechanisms (Chapter 4), resulting in a slower regeneration of cellular $NAD^+$ pools compared with the anaerobic mechanism in which lactate

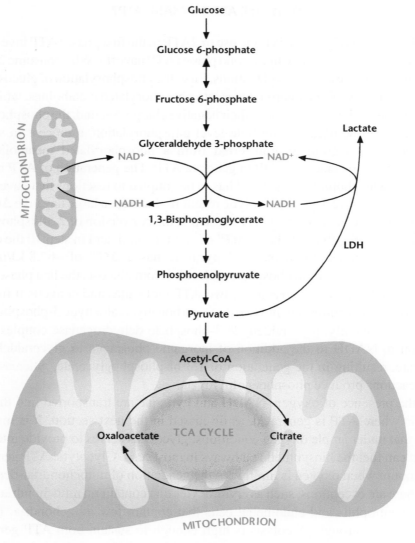

**Figure 3-4.** Regeneration of $NAD^+$ by LDH or mitochondria.

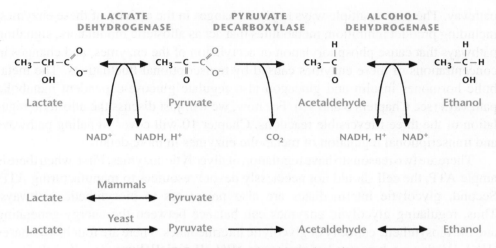

**Figure 3-5.** Fermentation in mammals and yeast.

dehydrogenase (LDH) reduces pyruvate into lactate by coupling this reaction to NADH oxidation to $NAD^+$. Many of us experience a buildup of lactate upon vigorous exercise. Under these conditions, oxygen cannot be delivered fast enough to muscle mitochondria to keep up with the high metabolic demands placed on muscle cells. In the absence of oxygen, the electron transport chain cannot regenerate $NAD^+$, making the LDH reaction critical for $NAD^+$ regeneration and continuous flux through glycolysis. This is sometimes referred to as anaerobic glycolysis or homolactic fermentation. The production of lactic acid is one form of fermentation using pyruvate. The other form of fermentation is the one most of us think of when we use yeast to make beer. This two-step process regenerates $NAD^+$ with two essential products required for beer, $CO_2$ and ethanol (Fig. 3-5). Yeast make greater amounts of ethanol through glucose consumption under anaerobic conditions compared with aerobic conditions, a phenomenon first observed by Louis Pasteur. This slowing down of fermentation in the presence of oxygen is referred to as the Pasteur effect. The opposite effect, in which the presence of high levels of glucose slows mitochondria from generating ATP under aerobic conditions, is called the Crabtree effect. Although the molecular details underlying the Pasteur and Crabtree effects are not fully understood, the short-term regulation of the Pasteur and Crabtree effects is controlled by metabolites regulating glycolytic enzymes (see the next section), whereas the long-term regulation is controlled by transcription factors regulating expression of glycolytic enzymes (Chapter 10).

## WHAT ARE THE KEY STEPS OF REGULATION?

The three irreversible reactions of glycolysis catalyzed by hexokinase, phosphofructokinase-1, and pyruvate kinase are important regulatory nodes within the glycolytic pathway. There are multiple ways to elicit changes in the activity of these enzymes, including product inhibition, metabolites that act as allosteric modulators, signaling pathways that cause phosphorylation or acetylation of the enzymes, and changes in concentrations of these enzymes caused by transcriptional fluctuations. The metabolic hormones insulin and glucagon also regulate glucose-dependent metabolic pathways (see Chapters 6 and 10). For now, we will just discuss the allosteric regulation of the three irreversible reactions. Chapter 10 will cover signaling pathways and transcriptional regulation of metabolic enzymes in more detail.

There are two reasons to have regulation of glycolytic enzymes. First, when there is ample ATP, the cell should not needlessly devote resources to manufacturing ATP. Second, glycolytic intermediates are also precursors to biosynthetic pathways. Thus, regulating glycolytic enzymes can balance between the energy-generating and the biosynthetic capacity of glycolytic intermediates. There are four hexokinases (HKI–IV) that catalyze step 1 of glycolysis. HKI, -II, and -III have a low $K_m$ (<0.5 mM) for glucose and are inhibited by glucose 6-phosphate, which accumulates if glycolysis is inhibited downstream from this reaction, a process referred to as feedback inhibition. This regulatory step ensures that glucose and ATP (in reactions 1 and 3) are not committed to glycolysis unless necessary. In contrast, hexokinase IV (also called glucokinase) has a high $K_m$ (6 mM) for glucose and is not inhibited by glucose 6-phosphate. Glucokinase has a higher $V_{max}$ compared with the other hexokinases. Mean blood glucose levels in nondiabetic humans are typically ~5 mM, but can increase up to 8 mM after a carbohydrate-rich meal full of sugar molecules. Glucokinase, which is abundant in the liver, has a high $V_{max}$ and effectively removes excess glucose from blood to minimize hyperglycemia after eating. Because of glucokinase's high $K_m$, its enzymatic activity diminishes once the blood sugar levels decrease to <6 mM.

High levels of ATP allosterically inhibit reactions 3 and 10, which are catalyzed by phosphofructokinase-1 (PFK1) and pyruvate kinase, respectively (Fig. 3-6). In contrast, when cellular ATP usage increases, the resulting ADP is quickly converted into ATP and AMP by adenylate kinase, buffering against a dramatic decrease in ATP levels (see the Chapter 2 discussion about energy charge). AMP levels can increase drastically during high ATP usage. AMP overcomes the ATP inhibition of PFK1. Another powerful regulatory mechanism involves the activity of the enzyme phosphofructokinase-2, which produces fructose 2,6-bisphosphate from fructose 6-phosphate, an allosteric activator of PFK1. Fructose 2,6-bisphosphate decreases the inhibitory effects of ATP and inhibits fructose 1,6-bisphosphatase, an enzyme involved in gluconeogenesis that catalyzes the reversal of reaction 3. This ensures that glycolysis occurs over gluconeogenesis. The product of PFK1, fructose 1,6-bisphosphate, activates pyruvate kinase to ensure the concentration of metabolites is low between fructose 1,6-bisphosphate and pyruvate, thus making these reactions thermodynamically favorable in the forward direction. This is an example of a feed-forward activation mechanism; it coordinates the upstream and downstream reactions of glycolysis. TCA cycle intermediary metabolites can also regulate

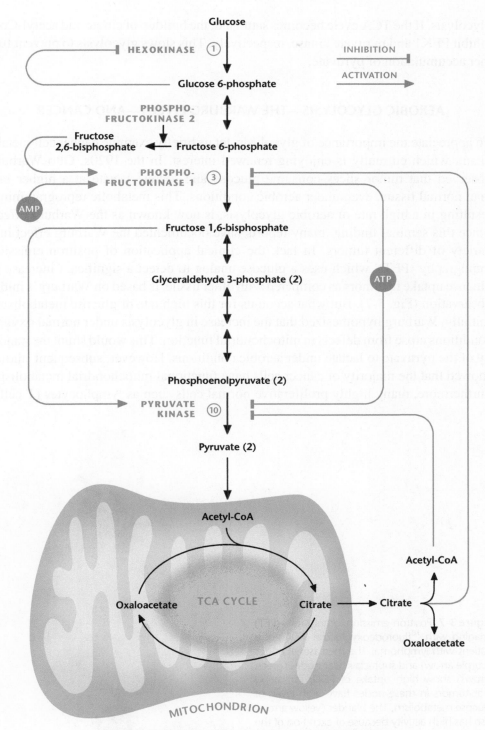

**Figure 3-6.** Regulation of glycolysis by metabolites.

glycolysis. If the TCA cycle becomes saturated, the buildup of citrate and acetyl-CoA inhibit PFK1 and pyruvate kinase, respectively. This slows glycolysis to prevent further accumulation of pyruvate.

## AEROBIC GLYCOLYSIS—THE WARBURG EFFECT—AND CANCER

To appreciate the importance of glycolysis, let us briefly examine cancer cell metabolism, which currently is enjoying renewed interest. In the 1920s, Otto Warburg observed that tumor slices consume glucose and secrete lactate at a higher rate than normal tissue, even under aerobic conditions. This metabolic reprogramming, resulting in a high rate of aerobic glycolysis, is now known as the Warburg effect. Since this seminal finding, many reports have documented the Warburg effect in a variety of different tumors. In fact, the clinical application of positron emission tomography (PET), which uses a glucose analog to detect a significant increase in glucose uptake in tumors as compared with other tissue, is based on Warburg's initial observation (Fig. 3-7). But what accounts for this high rate of glucose metabolism? Initially, Warburg hypothesized that the increase in glycolysis under normal oxygen conditions arose from defects in mitochondrial function. This would shunt the majority of the pyruvate to lactate under aerobic conditions. However, subsequent studies showed that the majority of cancer cells have functional mitochondrial metabolism. Furthermore, many highly proliferative normal cells such as lymphocytes (T cells)

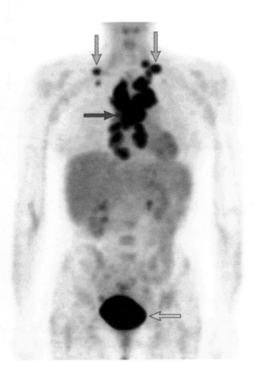

**Figure 3-7.** Positron-emission tomography (PET) imaging with $^{18}$fluorodeoxyglucose (FDG) of a patient with lymphoma. The mediastinal nodes (purple arrow) and supraclavicular nodes (green arrows) show high uptake of FDG, indicating that tumors in these nodes have high levels of glucose metabolism. The bladder (yellow arrow) also has high activity because of excretion of the radionuclide. (Reprinted from Gatenby and Gillies 2004, with permission from Macmillan Publishers.)

show the Warburg effect. As we will learn in Chapter 11, recent studies indicate that, under aerobic conditions, both normal proliferating cells and cancer cells activate signaling pathways and transcription factors to substantially increase the activity and expression of enzymes, respectively, to increase metabolic flux through glycolysis and the associated biosynthetic pathways. The advantage of the Warburg effect lies not necessarily in generating copious amounts of glycolytic ATP but in the biosynthetic capacity of glycolysis. Mitochondria are able to generate ample ATP in most proliferating cells, including cancer cells. The duplication of a cell into two daughter cells involves de novo synthesis of macromolecules such as lipids and nucleotides. Many of the glycolytic intermediates are precursors to anabolic pathways, including the pentose phosphate pathway (which generates NADPH and ribose 5-phosphate), the hexosamine pathway for glycosylation, and amino acid and lipid synthesis. These pathways will be discussed in more detail in later chapters.

## REFERENCE

Gatenby RA, Gillies RJ. 2004. Why do cancers have high aerobic glycolysis? *Nat Rev Cancer* **4:** 891–899.

## ADDITIONAL READING

Bar-Even A1, Flamholz A, Noor E, Milo R. 2012. Rethinking glycolysis: On the biochemical logic of metabolic pathways. *Nat Chem Biol* **8:** 509–517.

Crabtree HG. 1928. The carbohydrate metabolism of certain pathological overgrowths. *Biochem J* **22:** 1289–1298.

KEGG Pathway: Glycolysis/gluconeogenesis. http://www.genome.jp/kegg-bin/show_pathway?map 00010.

Koppenol WH, Bounds PL, Dang CV. 2011. Otto Warburg's contributions to current concepts of cancer metabolism. *Nat Rev Cancer* **11:** 325–337.

Locasale JW, Cantley LC. 2011. Metabolic flux and the regulation of mammalian cell growth. *Cell Metab* **14:** 443–451.

Lunt SY, Vander Heiden MG. 2011. Aerobic glycolysis: Meeting the metabolic requirements of cell proliferation. *Annu Rev Cell Dev Biol* **27:** 441–464.

Minakami S, de Verdier CH. 1976. Calorimetric study on human erythrocyte glycolysis heat production in various metabolic conditions. *Eur J Biochem* **65:** 451–460.

Nelson DL, Cox MM. 2013. *Lehninger principles of biochemistry*, 6th ed. WH Freeman, New York.

Warburg O. 1956. On the origin of cancer cells. *Science* **123:** 309–314.

# 4

# Mitochondria

I N HIS ROLLICKING 2005 BOOK, *Power, Sex and Suicide: Mitochondria and the Meaning of Life*, Nick Lane illustrates how mitochondria rule the biological world, something we mitochondrial biologists have always suspected. The title of the book references what mitochondria can do—they generate ATP (power), their DNA is maternally inherited (sex), and they invoke programmed cell death (suicide). The "meaning of life" alludes to the evolution of the first eukaryote. A current leading hypothesis, the endosymbiont theory, suggests that ~2 billion years ago two prokaryotes, an archaeon and an α-proteobacterium, developed a biological relationship in which both were mutually dependent on one another for nutritional requirements. Eventually, the archaeon, the host cell, acquired the α-proteobacterium, which became the primordial mitochondria. It is often assumed that the original symbiosis was based on the α-proteobacterium detoxifying oxygen and providing copious amount of ATP for the host cell. However, current data indicate that the symbiosis occurred before the presence of oxygen in the atmosphere, and it was the exchange of certain metabolites between the host archaeon and α-proteobacterium that was the metabolic basis of symbiosis. In fact, present-day mitochondria continue to constantly exchange metabolites with the cell. Interestingly, it is not clear whether the archaeon developed a nucleus before acquiring the α-proteobacterium. Furthermore, how the bacterium got inside the archaeon, the host cell, is still a mystery. But, like in many endosymbiotic relationships found in nature, the α-proteobacterium, over time, passed most of its DNA to the host nucleus, keeping just some of its genes. In fact, to date, all eukaryotic cells have mitochondria, or once had them and later lost them. *Rickettsia prowazekii*, the cause of typhus, has the most mitochondrial-like bacterial genome and bacteria, such as *Paracoccus denitrificans*, have similar biochemistry to a modern mitochondrion.

Today, we think of mitochondria as essential to maintenance of homeostasis in eukaryotes. If they fail to perform their critical functions, then pathologies result. A current leading hypothesis underlying the causes of diabetes, neurodegeneration, and aging is decline of mitochondria function (see the section Mitochondria and Disease). The biochemistry of mitochondria is well understood, but how this interfaces with the rest of the cell is not fully understood. Mitochondrial biologists hope to understand how these "living bacteria" within us work to shed insight into normal physiology and pathology.

## ESSENTIAL FEATURES OF EUKARYOTIC MITOCHONDRIA

1. Mitochondria are oval-shaped organelles that have five distinct parts. They contain an outer and inner membrane with a space between the membranes called the intermembrane space. The cristae are formed by the infoldings of the inner membrane, and the matrix is the space within the inner membrane. Mitochondria are dynamic organelles that undergo fusion and fission to form constantly changing tubular networks.

2. Mitochondria can use glycolysis-derived pyruvate, fatty acids, and amino acids to generate ATP through a process known as oxidative phosphorylation.

3. Mitochondria are biosynthetic hubs that produce metabolites, which can enter anabolic pathways to generate glycogen, lipids, nucleotides, and proteins. Mitochondria also produce heme and iron–sulfur clusters for certain proteins.

4. The nuclear DNA encodes most mitochondrial proteins. However, mitochondria have their own circular DNA, resembling bacterial plasmids, that encode for a subset of critical proteins. A mitochondrion can have multiple copies of its DNA. Mitochondrial DNA (mtDNA) is maternally inherited. mtDNA encodes 37 genes, including 13 critical for oxidative phosphorylation. The remaining approximately 3000 genes affecting mitochondria are encoded in the cell nucleus, and the resultant proteins are transported to the mitochondria.

5. Mitochondria are signaling organelles that regulate multiple biological processes, including proliferation, cell death, and metabolic adaptation. Mitochondria regulate these biological processes through multiple mechanisms including, but not limited to, release of ROS and metabolites, such as acetyl-CoA, calcium, cytochrome *c*, and mtDNA. Furthermore, the outer mitochondrial membrane serves as a platform for signaling by serving a scaffold for many critical proteins involved in cellular signaling.

## QUICK GUIDE TO THE ENERGY-GENERATING CAPACITY OF MITOCHONDRIA

Most mitochondrial metabolic activity occurs in the mitochondrial matrix either by soluble enzymes or protein parts of complexes embedded in the inner mitochondrial membrane.

- The inner mitochondrial membrane is impermeable to most ions and metabolites and has numerous transporters to shuttle ATP, pyruvate, and citrate. ATP and ADP are shuttled by the adenine nucleotide transporter (ANT). In contrast, the outer mitochondrial membrane is quite permeable to small metabolites and ions because of the presence of multiple copies of a porin, called a voltage-dependent anion channel (VDAC) (Fig. 4-1).

- The TCA cycle begins with the two-carbon acetyl-CoA and combines with four-carbon oxaloacetate (OAA) to generate citrate. The completion of the TCA cycle generates 3 NADH and 1 $FADH_2$.

- Acetyl-CoA can be generated from pyruvate by pyruvate dehydrogenase (PDH) or through fatty acid oxidation.

- Multiple amino acids can feed into the TCA cycle. Notably, glutamine can be converted into glutamate, which, subsequently, can generate the TCA cycle metabolite $\alpha$-ketoglutarate (2-oxoglutarate) in a process known as glutaminolysis.

- NADH and $FADH_2$ can feed the electron transport chain (ETC) complex I (NADH dehydrogenase) and complex II (succinate dehydrogenase, SDH), respectively, which pass their electrons to ubiquinone (Q). Complex III (ubiquinol–cytochrome *c* reductase) transfers electrons from Q to cytochrome *c*. Complex IV (cytochrome *c* oxidase) transfers electrons from cytochrome *c* to oxygen, forming $H_2O$. Complex I–III–IV and Complex II–III–IV form two different supercomplexes.

- Complexes I, III, and IV extrude $H^+$ from the matrix into the mitochondrial intermembrane space, thus, functioning as proton pumps. This generates proton motive force (pmf) composed of a small chemical component, $\Delta pH$, and a large electrical component, membrane potential $\Delta\psi$. Complex V ($F_1F_0$-ATP synthase) allows reentry of $H^+$ back into the matrix, allowing the phosphorylation of ADP to generate ATP by using the pmf. This process is coupled to the use of oxygen and is referred to as oxidative phosphorylation.

- Complex II participates in both the TCA cycle and ETC but does not conduct proton pumping.

## THE TCA CYCLE IS AN ENERGY-GENERATING PATHWAY

In 1937, Hans Krebs and his colleague William Johnson published a groundbreaking paper, "The role of citric acid in intermediate metabolism in animal tissues" in *Enzymologia*, outlining the citric acid cycle, also known as TCA cycle or Krebs cycle. Initially, Krebs submitted his results to *Nature*; however, the editors at the journal cited a backlog of papers and thus could not publish it without significant delay. In his memoir, Krebs wrote, "This was the first time in my career, after having published more than fifty papers, that I experienced a rejection or semi-rejection."

In 1988, seven years after Krebs's death, an anonymous editor published a letter in *Nature* acknowledging their mistake by stating "rejection of Hans Krebs'

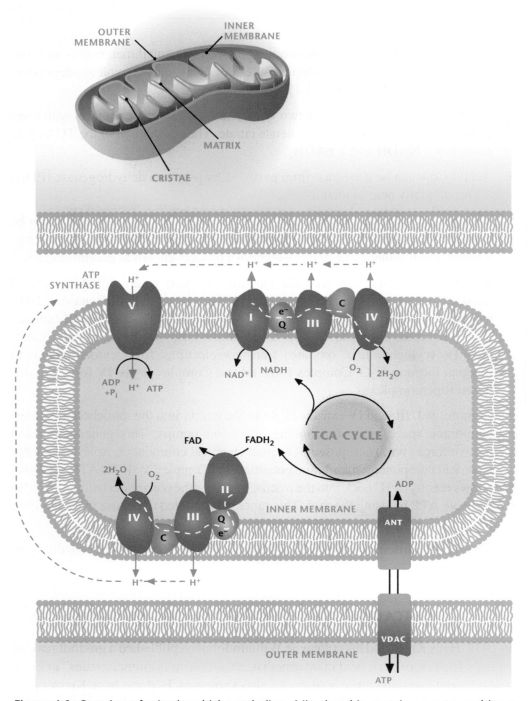

**Figure 4-1.** Overview of mitochondrial metabolism. Mitochondria contain an outer and inner membrane. The mitochondrial matrix enclosed by the inner membrane contains the TCA cycle that generates reducing equivalents NADH and $FADH_2$, which donate electrons to the electron transport chain (ETC), resulting in the generation of a proton motive force to drive ATP synthesis. The ETC is composed of two supercomplexes consisting of either complex I–III–IV or II–III–IV. (Modified from http://en.wikipedia.org/wiki/Electron_transport_chain#mediaviewer/ File:Mitochondrial_electron_transport_chain%E2%80%94Etc4.svg.)

discovery of the tricarboxylic (Krebs') cycle, a pivot of biochemical metabolism, remains *Nature's* most egregious error (as far as we know)."

The TCA cycle lies at the core of eukaryotic cell metabolism because multiple substrates can feed into the cycle, including fatty acids, amino acids, and pyruvate generated from glucose via glycolysis. The cycle is distinct from linear pathways, like glycolysis, because OAA is the substrate for the first reaction catalyzed by citrate synthase and the product of the last reaction catalyzed by malate dehydrogenase (MDH), thus renewing the cycle. The TCA cycle is identified as an amphibolic pathway because it can provide intermediates for macromolecules (e.g., lipids) and synthesis (i.e., anabolism) and can generate the reducing equivalents NADH and FADH$_2$ to ultimately produce ATP through oxidative phosphorylation (i.e., catabolism).

The TCA cycle begins when acetyl-CoA is generated by oxidation of pyruvate (e.g., from glycolysis; see Chapter 3) or other compounds, including fatty acids (Chapter 7) and amino acids (Chapter 8). The massive PDH complex converts three-carbon pyruvate to a two-carbon acetyl-CoA. The primary function of one turn of the TCA cycle from an energy-generating perspective is to oxidize acetyl-CoA to two CO$_2$ molecules. The released electrons are transferred to the coenzymes NAD$^+$ and FAD to form three NADH and one FADH$_2$ molecules, which are reoxidized and used by the ETC to generate ATP through oxidative phosphorylation, discussed subsequently in this chapter.

In the first of eight reactions of the TCA cycle, the two-carbon acetyl-CoA transfers the acetyl group to the four-carbon OAA to generate a six-carbon citrate (Fig. 4-2). The cycle proceeds with two oxidative decarboxylation in steps 3 and 4, resulting in release of two molecules of CO$_2$ and generation of two NADH molecules and a four-carbon succinyl-CoA. The fifth step of the TCA cycle converts succinyl-CoA into succinate, coupling it to the generation of GTP, which can be converted into ATP. The remaining three steps of the TCA cycle convert four-carbon succinate into a four-carbon OAA that can combine with another acetyl-CoA molecule to continue the TCA cycle. Most of the reactions of the TCA cycle are reversible except for the reactions catalyzed by citrate synthase, isocitrate dehydrogenase, and α-ketoglutarate dehydrogenase, which have the largest negative $\Delta G^{\circ\prime}$. In contrast, the last reaction catalyzed by MDH is the most unfavorable reaction and thus requires energy ($\Delta G^{\circ\prime} = +29.7$ kJ/mole). Remember from our previous discussion in Chapter 2 that keeping the product of a reaction very low compared with reactant (law of mass action) can make an unfavorable reaction favorable. Thus, the OAA concentration in the matrix of the mitochondria is kept extremely low, allowing the last reaction to have a favorable negative actual Gibbs free energy.

The overall reaction of the TCA cycle is

$$\text{Acetyl-CoA} + 3\text{NAD}^+ + 1\text{FAD} + 1\text{GDP} + P_i + \rightarrow 1\text{CoA-SH} + 2\text{CO}_2$$
$$+ 3\text{NADH} + 1\text{FADH}_2 + 1\text{GTP}, \ \Delta G^{\circ\prime} = -57.3 \text{ kJ/mol}.$$

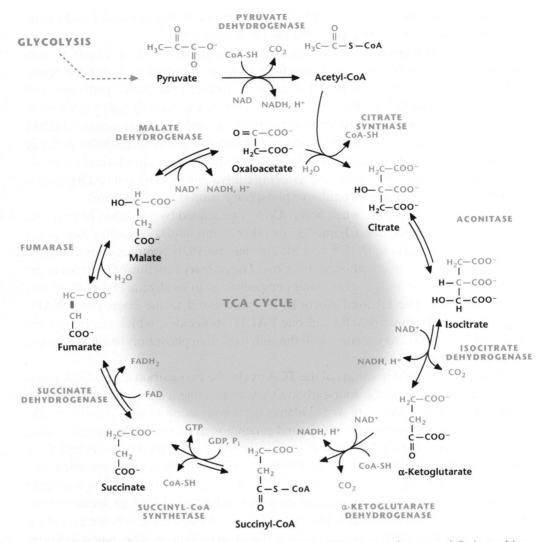

**Figure 4-2.** Overview of TCA cycle. The TCA cycle starts when two-carbon acetyl-CoA combine with four-carbon OAA to generate citrate. The TCA cycle produces three NADH, one $FADH_2$, and one GTP.

## THE TCA CYCLE IS A BIOSYNTHETIC HUB

In his Nobel lecture in 1953, Hans Krebs noted "in some micro-organisms the cycle primarily supplies intermediates rather than energy, whilst in the animal and most other organisms it supplies both energy and intermediates." Many of the enzymes involved in the TCA cycle were abundant well before the presence of oxygen in the Earth's environment. The likely function of the earliest components of the

TCA cycle was to provide metabolites that could be used for biosynthesis. In eukaryotes, notable biosynthetic reactions in the TCA cycle are as follows (Fig. 4-3).

1. OAA is converted to phosphoenolpyruvate, which is a substrate for gluconeogenesis.

2. α-Ketoglutarate, also known as 2-oxoglutarate, is converted to glutamate, which can generate glutamine for use as a precursor to generate purine nucleotides (adenosine and guanosine).

3. Succinyl-CoA is a precursor of prophyrins, such as heme.

4. Citrate is exported to the cytosol where it is converted into OAA and acetyl-CoA. OAA can generate aspartate, which is used for purine and pyrimidine synthesis. Acetyl-CoA is used for lipid synthesis.

An important consideration is that when TCA-cycle intermediates are being siphoned off for biosynthetic purposes, the cycle has to be replenished or else it would cease. The replenishment of the TCA cycle is referred to as anaplerosis, which

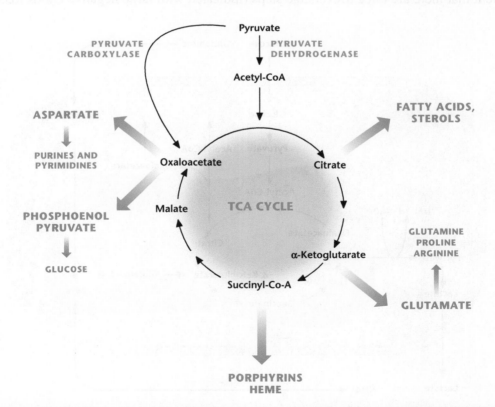

**Figure 4-3.** The TCA cycle is a biosynthetic hub. TCA-cycle intermediates provided the building blocks for macromolecules including fatty acids, nucleotides, hemes, and porphyrins. Oxaloacetate provides the carbons for the generation for glucose (i.e., gluconeogenesis).

means "filling up." As we will learn in Chapter 6, the initial steps of gluconeogenesis involve the conversion of OAA to phosphoenolypyruvate, which goes through a series of reactions to make glucose. Thus, if OAA is being depleted, it has to be replenished to maintain a minimal level to allow the cycle to function. There are multiple inputs into the TCA cycle, but two important anaplerotic mechanisms are the conversion of pyruvate to mitochondrial OAA by pyruvate carboxylase and the conversion of glutamine to glutamate and subsequently to α-ketoglutarate, referred to as glutaminolysis (Fig. 4-4). The latter mechanism is often used when citrate is exported from the mitochondria into the cytosol for de novo lipid synthesis, thus preventing the formation of subsequent TCA-cycle intermediates, such as α-ketoglutarate. Glutaminolysis feeds α-ketoglutarate into the TCA cycle, thereby allowing the cycle to continue functioning.

## REGULATION OF THE TCA CYCLE

If we examine the changes in free energy of the reactions of the TCA cycle, it is apparent that there are three irreversible steps (indicated with large negative Gibbs free-

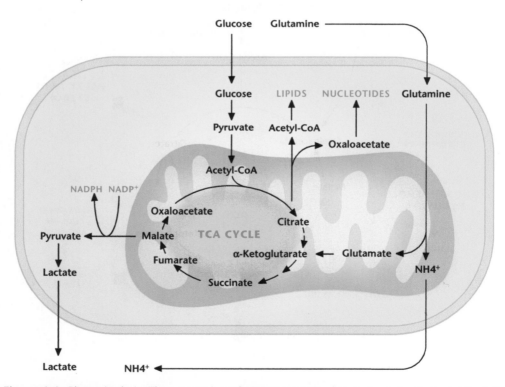

**Figure 4-4.** Glutaminolysis. The conversion of glutamine into glutamate and, subsequently, into α-ketoglutarate is referred to as glutaminolysis. This is an important pathway to replenish TCA-cycle intermediates.

energy values) that are catalyzed by citrate synthase, isocitrate dehydrogenase, and α-ketoglutarate dehydrogenase, but it should be noted that isocitrate dehydrogenase can be reversible under certain conditions. A fourth irreversible reaction that is not technically part of the TCA cycle, but is an important regulator, is PDH, which generates acetyl-CoA from pyruvate. The TCA cycle is regulated by substrate availability, such as the levels of OAA and acetyl-CoA that are required to initiate the TCA cycle. There are multiple positive and negative allosteric regulators that control the metabolic flux of the TCA cycle, including acetyl-CoA, succinyl-CoA, ATP, ADP, AMP, $NAD^+$, and NADH (Fig. 4-5). NADH inhibits all of the regulatory enzymes in the TCA cycle. Because NADH generates ATP through the ETC and oxidative phosphorylation, ATP also is an allosteric inhibitor of PDH and isocitrate dehydrogenase. Thus, when cells have ample NADH and ATP, the cycle slows down. In contrast, high demand for ATP increases the ADP/ATP ratio and AMP levels, resulting in stimulation of the regulatory enzymes of the TCA cycle (Fig. 4-5).

Three metabolites that are important regulators of the TCA cycle are acetyl-CoA, succinyl-CoA, and OAA. Abundant acetyl-CoA inhibits PDH but activates pyruvate carboxylase to shunt pyruvate to OAA, thus balancing the entry of OAA with acetyl-CoA. Succinyl-CoA inhibits both citrate synthase and α-ketoglutarate dehydrogenase to slow the cycle down. Citrate generated from acetyl-CoA and OAA is diminished and α-ketoglutarate generated by glutaminolysis is also decreased. An increase in OAA can slow the cycle down by inhibiting SDH. Finally, calcium is a positive regulator of the cycle by activating pyruvate, isocitrate, and α-ketoglutarate

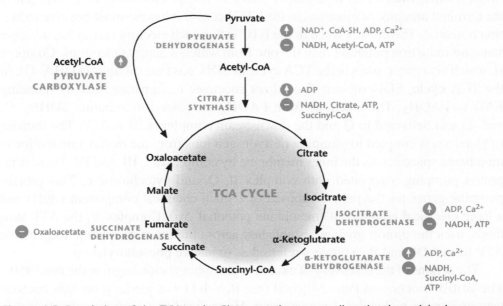

**Figure 4-5.** Regulation of the TCA cycle. Citrate synthase, as well as the three dehydrogenases, isocitrate, α-ketoglutarate, and PDH, are stimulated when ADP levels increase and dampened when NADH levels increase. The three dehydrogenases are positively regulated by calcium.

dehydrogenase. This is an important mechanism in muscle cells that use calcium for contraction, an energy-demanding process. The increase in cytoplasmic calcium triggers uptake of calcium in the mitochondria, and this stimulates these dehydrogenases to couple muscle contraction to the energy machinery. An exciting new area involves the multiple cellular signaling mechanisms uncovered recently that regulate the TCA cycle, as we will discuss in Chapter 10.

## BASIC ASPECTS OF OXIDATIVE PHOSPHORYLATION

The reducing equivalents NADH and $FADH_2$ generated by the TCA cycle have to be oxidized to $NAD^+$ and FAD so that the cycle can continue to function. The ETC, also called the respiratory chain, consists of complexes I, II, III, and IV embedded in the inner membrane of the mitochondria (Fig. 4-6). Complex I, also referred to as NADH dehydrogenase, is composed of 46 subunits in human cells. It oxidizes NADH to $NAD^+$ and passes two electrons to Q. Subsequently, Q donates two electrons to complex III and becomes ubiquinol (QH2). Complex III is also called the bc1 complex, which is composed of 11 subunits. Q then becomes ubiquinol. Complex III passes one electron at a time to cytochrome $c$, which donates electrons to complex IV, also known as cytochrome $c$ oxidase, which is composed of 13 subunits. Complex IV takes four electrons from cytochrome $c$ and donates them to the final electron acceptor molecular oxygen to generate two molecules of water ($O_2 + 4e + 4H \rightarrow 2H_2O$). Remember from our discussion in Chapter 2 that a molecule with a high-(most-positive) reduction potential is likely to accept electrons. Thus, oxygen is the terminal acceptor of electrons in the ETC because it has the most-positive reduction potential. The driving force of the ETC is that each electron carrier has a higher standard reduction potential than the one from which it accepts electrons. Complex II, which also participates in the TCA cycle as SDH, has four subunits (SDHA–D). In the TCA cycle, SDH subunit A oxidizes succinate to fumarate, thereby reducing FAD to $FADH_2$. The electrons from $FADH_2$ are passed to subunits SDHB, -C, and -D and delivered to Q and the downstream complexes III and IV. The transfer of electrons is coupled to pumping of hydrogen ions from the matrix into the inter-membrane space across the inner membrane by complexes I, III, and IV. There is no proton pumping associated with complex II, Q, and cytochrome $c$. This protein pumping generates the pmf, composed of a small chemical component ($\Delta pH$) and a large electrical component (membrane potential $\Delta\psi$). Complex V, the ATP synthase, uses the proton gradient established across the inner membrane to generate ATP from ADP and $P_i$, a process referred as oxidative phosphorylation.

The origins of the discovery of oxidative phosphorylation begin in the late 1950s. The British biochemist Peter Mitchell (see Box 4-1) was working on how bacteria take up certain molecules, such as sugar. He realized that proton gradients across the bacterial membranes were critical for nutrient uptake, and this led to his incredible insight that ATP synthesis is powered by proton gradients. Proton gradients are

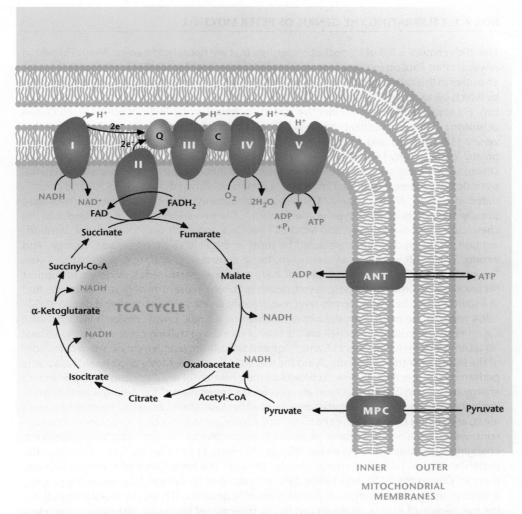

**Figure 4-6.** Oxidative phosphorylation. The electron transport chain complex elements I, III, and IV pump protons across the inner mitochondrial membrane as electrons are transferred through the chain to oxygen, the final electron acceptor. This protein pumping generates the pmf, composed of a small chemical component ($\Delta$pH) and a large electrical component (membrane potential, $\Delta\psi$). Complex V, the ATP synthase, uses the proton gradient established across the inner membrane to generate ATP from ADP and $P_i$, a process referred as oxidative phosphorylation.

ubiquitous in nature and are essential for sustaining life. Proton power drives photosynthesis, active transport of molecules in an out of cells, and ATP generation. Even the earliest forms of life, such as $\alpha$-proteobacteria and archaea, had proton pumps to generate ATP. This coupling of ATP synthesis to proton pumping, known as the "chemiosmotic theory," is one of the key discoveries of the 20th century.

The salient feature of the chemiosmotic theory is that two systems, the transfer of electrons through the ETC and generation of ATP through complex V, are related

---

**BOX 4-1. CELEBRATING THE GENIUS OF PETER MITCHELL**

---

The 20th century is full of iconoclastic scientists that are household names: Albert Einstein in physics, Linus Pauling in chemistry, and James Watson and Francis Crick in biology. Perhaps an outlier in this category is Peter Mitchell (1920–1992), the man who discovered the process by which living organisms generate energy. His career was unique compared with most scientists. His ideas stemmed from inductive reasoning, based on his findings on the essential role of proton pumping in nutrient uptake in bacteria during the years he spent at Cambridge University (1939–1955) and Edinburgh University (1955–1963). During these years, he was greatly influenced by Frederick Gowland Hopkins, James Danielli, David Keilin, and Malcolm Dixon, as well as the ancient Greek philosophers. The theoretical ideas that became his chemiosmotic theory were introduced in a series of papers published during the 1950s and consolidated in a 1961 *Nature* paper, "Coupling of phosphorylation to electron and hydrogen transfer by a chemiosmotic type of mechanism." The prevailing theory at the time was the chemical intermediate theory, and Mitchell's vectorial approach with ideas of pmf, symport, antiport, and uniport were considered by most in the field to be retrograde, strange, and wrong. The result was years of skepticism, fierce controversy, and heated exchanges—the "ox-phos" wars. Initially Mitchell did not have extensive experimental evidence to support his revolutionary ideas. But oddly, severe health problems in the form of gastric ulcers resulted in a solution. Mitchell, using family resources, bought an estate in Cornwall and during 1962–1964 restored the estate with the aim of establishing a small, private research institute—the Glynn Research Institute. Mitchell credits several months of milking cows as helping to heal his ulcers. His research partner at Cambridge and Edinburgh, Jennifer Moyle, was a cofounder of the Institute. With Mitchell supplying the theoretical analysis, Moyle and research assistants performed the specific experiments providing the evidence for the chemiosmotic approach to energy generation in the cell. Glynn also provided a "quiet haven" for visiting scientists working in this burgeoning bioenergetics area. Soon other laboratories began to confirm Mitchell's ideas, and during the 1970s the chemiosmotic theory gained reluctant acceptance, although controversy remained for many years over proton/electron ratios. Mitchell's paradigm-changing ideas resulted in the Nobel Prize in Chemistry in 1978 for his "contribution to the understanding of biological energy transfer through the formulation of the chemiosmotic theory." During the 1970s and 1980s, Mitchell extended his theories to propose the Q cycle, a mechanism by which complex III pumps protons to generate ATP, and to develop a model for the mechanism of ATPase. As evidenced by the trajectory of his career, he believed that scientists should be free of bureaucratic interference, as well as social responsibility; it was up to society to determine whether scientific results were to be used for good or ill. The Glynn Research Institute continued its work after Mitchell's death, but it was difficult to raise sufficient funds to continue its high level of investigation. In 1996, the Institute was absorbed into the University College of London as the Glynn Laboratory of Bioenergetics.

through a pmf. We are used to thinking that when one particular process exerts influence on another, there are usually protein–protein interactions. But here the two systems are not in close contact or proximity and they communicate through the pmf generated by pumping of protons from the mitochondrial matrix into the intermembrane space through the inner membrane by complexes I, III, and IV. The vectorial $H^+$ pumping by complexes I, III, and IV generates both a chemical gradient, measured as a $\Delta pH$, and an electrical gradient based on separation of charge, measured as a membrane potential ($\psi$). The major contributor of the pmf is $\psi$, whereas the $\Delta$pH across the mitochondrial inner membrane is only 1 pH unit. The stoichiometry of proton pumping is $4H^+$ for complexes I and III but only $2H^+$ in complex IV. This results in $10H^+$ for one molecule of NADH oxidized and $6H^+$ for $FADH_2$ oxidized by the ETC. $FADH_2$ oxidation does not use complex I; thus, it only has $6H^+$ protons pumped through complexes III and IV. The protons flow back down the gradient through the inner membrane-bound complex V in response to the different chemical ($H^+$ concentration) and electrical (separation of charge) potentials across the membrane. We can think of this as a proton circuit.

Complex V is composed of two distinct multisubunit components, an ATP-hydrolyzing catalytic subunit known as the spherical "head" ($F_1$-ATPase), which is hydrophilic and protrudes into the aqueous mitochondrial matrix, and the hydrophobic proton-pumping component ($F_0$), embedded in the inner mitochondrial membrane. An isolated $F_1$-ATPase catalyzes the hydrolysis of ATP to ADP and $P_i$, which is why it is called the $F_1$-ATPase. However, in an intact mitochondrion, the protons that have accumulated in the mitochondrial intermembrane space as a result of ETC proton pumping enter the $F_0$ complex and exit into the matrix. The energy dissipated as the protons travel down their concentration gradient rotates $F_0$ in a clockwise direction, inducing repeated conformational changes in the $F_1$-ATPase and enabling the conversion of ADP and $P_i$ into ATP. This makes the complex V or $F_0F_1$-ATPase the smallest rotary machine ever known!

Mitochondrial oxygen consumption is a way to assess ETC and complex V function (see Box 4-2). Oxygen consumption is reflective of the reduction of oxygen to water by complex IV. The transfer of electrons through the ETC is coupled to pumping protons to generate the pmf. At some point, this force is high enough that the ETC cannot pump against this gradient anymore; when this happens, the transfer of electrons ceases and so does oxygen consumption. In the presence of ADP, the proton gradient can be dissipated through complex V, allowing the transfer of electrons through the ETC to complex IV, and oxygen consumption resumes. The rate of oxygen consumption by complex IV coupled to the generation of ATP synthesis by complex V is called coupled respiration. However, a portion of the protons can leak back across the inner membrane rather than moving to complex V. In the absence of ADP, this proton leak allows some dissipation of the proton gradient to permit electron transfer through the ETC, resulting in a small amount of oxygen consumption. The rate of oxygen consumption by complex IV that is not coupled to the generation of ATP, but is caused by the proton leak, is called uncoupled respiration. Thus, mitochondrial oxygen consumption is a combination of coupled respiration and uncoupled respiration. Most cells display high levels of coupled respiration. Brown fat cells are a notable exception. They show uncoupled respiration due to an abundance of uncoupling proteins, which increase proton leak by allowing protons to flow back into mitochondria without driving the generation of ATP. There are protonophores, such as carbonyl cyanide-*p*-trifluoromethoxyphenylhydrazone (FCCP)

---

**BOX 4-2. MEASURING THE CELLULAR OXYGEN CONSUMPTION RATE**

A tool to assess mitochondrial ETC function is to measure the cellular oxygen consumption rate (OCR) in the presence of various inhibitors. Typically, cellular OCR is measured by adding (1) the ATP synthase inhibitor oligomycin, (2) uncoupler FCCP, and (3) complex I and III inhibitors rotenone and antimycin, respectively. Basal respiration defined as mitochondrial OCR is calculated by subtracting the residual OCR after administering ETC inhibitors rotenone and antimycin from the cellular oxygen consumption (Box 4-2, Fig. 1). The residual OCR is referred to as nonmitochondrial respiration. Coupled respiration is calculated by subtracting the residual respiration upon the addition of oligomycin, referred to as uncoupled respiration, from basal respiration. Uncoupled respiration is a measurement of the proton leak. Maximal respiration can be calculated by the addition of FCCP, a potent protonophore, which uncouples mitochondrial ATP generation from oxygen consumption (Box 4-2, Fig. 1).

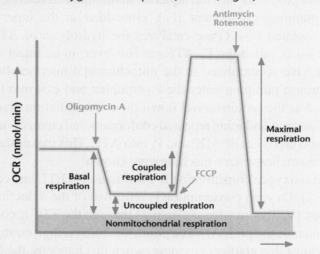

**Box 4-2, Figure 1.** Determining the oxygen consumption rate. (Modified from Anso et al. 2013.)

---

and 2,4-dinitrophenol (DNP), which allow protons to equilibrate across the inner mitochondrial membrane by increasing the proton leak. Thus, the proton gradient never builds up to slow the ETC. This allows transfer of electrons through the ETC to complex IV at a maximal rate, a process called maximal respiration. This rate will be dependent on how quickly NADH and $FADH_2$ can feed electrons to the ETC, as well as the efficiency of individual complexes to transfer electrons. Interestingly, in the 1930s DNP was used as an agent to combat obesity. However, it was quite toxic and quickly removed from the market because of an excessive increase in body temperature as a result of uncoupling. A recent study used a modified version of DNP targeted to the liver to decrease features of type II diabetes in rats, such as hyperglycemia, fatty liver, and insulin resistance. It will be of interest to see if this finding translates into humans.

It is important to note when the ETC is not pumping protons and no membrane potential or pH gradient is generated, for example, in the absence of oxygen, or if proteins of the ETC are damaged or mutated, then glycolysis-generated ATP is imported into mitochondria where the $F_1$-ATPase hydrolyzes this ATP to ADP and $P_i$, and protons flow from the matrix into the intermembrane space through $F_0$. The maintenance of a membrane potential is vital for protein import and export from the mitochondrial matrix. Important sets of proteins synthesized in the mitochondria are those that contain iron–sulfur clusters and are exported into the cytosol; here these proteins serve functions, such as maintenance of nuclear genomic integrity. If mitochondria lose their inner mitochondrial membrane potential, then they are tagged for degradation through a process known as mitophagy. There are proteins, such as PTEN-induced putative kinase 1, that recognize depolarized mitochondria and recruit Parkin, which coats mitochondria that have low-mitochondrial-membrane potential with ubiquitin, thereby targeting them for degradation. If this mechanism to clear dysfunctional mitochondria fails, then there is a potential to accumulate these low-membrane-potential mitochondria that do not import or export protein efficiently and have diminished biosynthetic capacity and ATP generation. Thus, Mitchell's chemiosmotic theory of proton pumping is essential for mitochondria to maintain their function.

Let us crunch the numbers to figure out how much ATP is generated by NADH and FADH$_2$ oxidation by the ETC. This is referred to as currency exchange. Three H$^+$ are required to synthesize 1ATP when they flow back down the electrochemical proton gradient through complex V, and 1H$^+$ is required to transport each negatively charged P$_i$ molecule into the matrix. Based on these numbers, 1NADH oxidation by complex I results in 10H$^+$ pumped out of the matrix by complexes I, III, and IV, and 4H$^+$ pumped in by complex V for ATP synthesis, resulting in 10H$^+$/4H$^+$ = 2.5ATP/NADH. FADH$_2$ oxidation by complex II generates only 6H$^+$ through complexes III and IV proton pumping, resulting in 6H$^+$/4H$^+$ = 1.5ATP/FADH$_2$.

## THE ETC GENERATES ROS

An unfortunate consequence of electron transfer through the ETC is the generation of $O_2^-$. The ETC is not a perfect system in which all the electrons from NADH or FADH$_2$ are transferred eventually to complex IV and oxygen, which is the final acceptor of these electrons. There is small leakage (estimated to be <0.1%) through the system in which the electrons react in a nonenzymatic manner with oxygen to generate $O_2^-$, which is produced by a one-electron reduction of $O_2$. $O_2^-$ is rapidly converted into $H_2O_2$ by the enzyme SOD2 in the mitochondrial matrix. There are multiple proteins, such as GSH peroxidases and peroxiredoxins, which detoxify $H_2O_2$ to $H_2O$ in the mitochondrial matrix (discussed in Chapter 5). If the ETC complexes are not functioning properly, the levels of $O_2^-$ and peroxide generation reach levels that can incur damage to mitochondrial proteins and mtDNA. Furthermore, $H_2O_2$ can leak from mitochondria, and at high levels this release can damage proteins in the cytosol, as well as oxidize guanine in the nuclear DNA. Abundant antioxidant protein levels limit the levels of ROS in the mitochondrial matrix. Based on isolated mitochondria studies, $O_2^-$ production increases when the mitochondria are not making ATP and, consequently, have a high pmf as a result of limiting ADP availability in the matrix. Superoxide production also increases when there is a high $NADH/NAD^+$ ratio in the mitochondrial matrix because of ETC inhibition. If mitochondria are actively generating ATP, then they have a lower pmf and $NADH/NAD^+$ ratio, resulting in lower production of $O_2^-$. A paradoxical observation in cells is that decreasing or increasing oxygen levels both stimulate $O_2^-$ production.

There are two unresolved questions regarding the efficiency of electron transfer through the ETC: How is it so efficient to begin with (i.e., there is only <0.1% leakage that generates $O_2^-$)? And, why has not nature selected to get rid of these "toxic molecules"? The answer to the first question might be that complexes I, III, and IV are found in distinct supercomplexes to allow for electrons to be channeled over short distances. This close proximity would increase efficiency, allowing tight coupling of electron transfer to proton pumping for the generation of membrane potential, as well as limiting electron leakage that would generate $O_2^-$. Recently, proteins such as hypoxia-inducible gene-1 (HIG1) have been identified as critical to the formation of these supercomplexes. The two mobile carriers Q and cytochrome $c$, which are not imbedded in the membranes, also seem to have distinct pools that are either freely mobile in the intermembrane space or bound close to the inner membrane in the intermembrane space. The answer to the second question could be that the low-level generation and release of ROS by mitochondria have a beneficial function. An emerging hypothesis is that the low levels of ROS emerging from mitochondria could serve as a mode of communication to the rest of the cell to convey its fitness. Thus, as healthy mitochondria generate low levels of ROS, their release in the cytosol targets specific ROS-sensitive proteins that are critical in maintaining homeostasis, as well as for adapting to stressful conditions. Recent studies indicate that mitochondrial-generated ROS leaking into the cytosol are necessary for optimal activation of hypoxic induction of genes for metabolic adaptation, autophagy, and innate and adaptive immunity. Furthermore, mitochondrial ROS have been shown to be beneficial in extending longevity in model organisms—a radical idea! (See Box 5-2.)

## TRANSPORTERS THAT MOVE METABOLITES IN AND OUT OF MITOCHONDRIA

So far, we have covered how coupling of complexes of the ETC to complex V generates ATP by pmf. An important consideration recognized in Mitchell's original paper as an essential component of the proton circuit is the transport of metabolites, such as ADP and $P_i$ into the mitochondria and transport of ATP into the cytosol. The outer mitochondrial membrane contains VDACs that allow these metabolites to be transported across this membrane. However, the inner mitochondrial membrane is impermeable to these metabolites and ions; otherwise, protons pumped by the ETC would flow right back and not go through complex V to generate ATP. The inner mitochondrial membrane contains an ANT, which catalyzes the exchange of ATP for ADP. ATP has four negative charges, whereas ADP has three negative charges at neutral pH, resulting in the loss of a net charge of −1 in the matrix. This transport uses some of the membrane potential generated by the ETC and favors the entry of ADP and exit of ATP. The phosphate transporter exchanges $H_2PO_4^-$ at neutral pH for $OH^-$ or cotransports with $H^+$. The phosphate enters via the pH gradient and not by the membrane potential generated by ETC. Thus, ATP synthesis requires four protons, three for ATP synthase and one for $P_i$ transport.

Chapter 3 covered how the glucose molecule becomes two pyruvate molecules through a series of metabolic reactions known as glycolysis. The two main products of glycolysis are NADH and pyruvate. If oxygen is present, then pyruvate and NADH are shuttled into the mitochondria. Both can pass through the outer mitochondrial membrane, which is permeable to small metabolites, but they need specialized mechanisms to cross the impermeable inner mitochondrial membrane. The elusive nature of the mitochondrial pyruvate carrier was deciphered recently. This carrier is composed of two proteins, MPC1 and MPC2, each ~15 kDa in size, that form a large heterocomplex. An important detail of this transporter is that it costs one proton to transport a molecule of pyruvate into the matrix. The molecular details of how this heterocomplex functions are currently being unraveled.

The NADH generated by glycolysis uses two shuttle mechanisms for transport into the mitochondrial matrix: the malate–aspartate and the glycerol 3-phosphate shuttles. The key enzyme in the malate–aspartate shuttle is MDH, located in the cytosol that oxidizes NADH to $NAD^+$ to allow glycolysis to function, whereas the mitochondrial matrix MDH reduces $NAD^+$ to NADH (Fig. 4-7). The glycerol 3-phosphate shuttle depends on the activity of two different glycerol 3-phosphate dehydrogenase enzymes, one using a NAD/NADH couple and the other a FAD-linked membrane-bound enzyme, which generates $FADH_2$ that donates electrons to Q of the ETC (Fig. 4-8).

The other major transporters for citrate, isocitrate, malate, succinate, and α-ketoglutarate were quickly demonstrated (Fig. 4-9). These are all important transporters, but let us pay close attention to the citrate transporter (SLC25A1), which allows TCA-cycle-generated citrate to cross the inner mitochondrial membrane into the intermembrane space, where citrate goes through the VDACs located in the outer mitochondrial membrane. Citrate$^{2-}$ is exchanged for malate$^{2-}$. Once in the cytosol, the six-carbon citrate can become two-carbon acetyl-CoA and four-carbon OAA by ATP citrate lyase (ACLY). As we discussed earlier, acetyl-CoA is an important metabolite that can be used for protein acetylation and fatty acid and sterol synthesis. The production of acetyl-CoA occurs predominantly in the mitochondria matrix. The

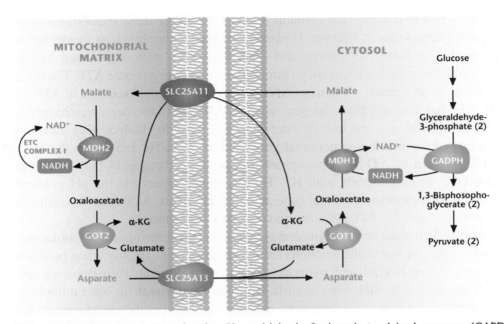

**Figure 4-7.** Malate–aspartate shuttle. Glyceraldehyde 3-phosphate dehydrogenase (GAPDH) generates NADH during glycolysis. NADH can be regenerated to NAD$^+$ to allow glycolysis to continue by the conversion of oxaloacetate (OAA) to malate by cytosolic malate dehydrogenase 1 (MDH1). Subsequently, malate is transported by the SLC25A1 transporter into the mitochondrial matrix, in which it is converted back to OAA coupled with NAD$^+$ conversion into NADH. The ETC complex I converts NADH into NAD$^+$ to keep malate dehydrogenase 2 functioning continuously. The mitochondrial OAA is converted into aspartate by aspartate aminotransferase 2 (GOT2) and, subsequently, transported into the cytosol, where aspartate can be converted back into cytosolic OAA by aspartate aminotransferase 1 (GOT1) for the shuttle to continue. α-KG, α-ketoglutarate. (Adapted with permission of themedicalbiochemistrypage, LLC.)

three main sources of acetyl-CoA are pyruvate oxidation by PDH, the breakdown of fatty acids (Chapter 7), and catabolism of amino acids. The mitochondrial membrane is impermeable to acetyl-CoA molecules, and so the citrate transporter is a major mechanism for delivering acetyl-CoA into the cytosol.

## COMPLETE OXIDATION OF GLUCOSE BY GLYCOLYSIS AND MITOCHONDRIA GENERATES 32 ATP

Using information from Chapter 3 on glycolysis and this chapter on the TCA and oxidative phosphorylation, we can now calculate how much ATP one molecule of glucose can generate (Fig. 4-10). From the combination of glycolysis, PDH reaction, and reducing equivalents generated by the TCA cycle, we obtain the net reaction

$$\text{Glucose} + 2H_2O + 10NAD^+ + 2FAD + 4ADP + 4P_i$$
$$\rightarrow 6CO_2 + 10NADH + 6H^+ + 2FADH_2 + 4ATP.$$

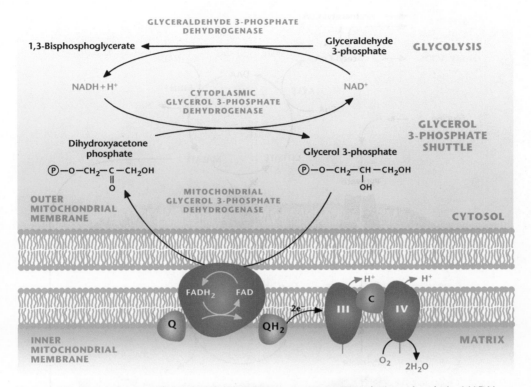

**Figure 4-8.** Glycerol–phosphate shuttle. GAPDH generates NADH during glycolysis. NADH can be regenerated to $NAD^+$ to allow glycolysis to continue by the conversion of dihydroxyacetone phosphate into glycerol 3-phosphate by cytosolic glycerol 3-phosphate dehydrogenase. Subsequently, glycerol 3-phosphate is converted back into dihydroxyacetone phosphate by coupling FAD into $FADH_2$. $FADH_2$ donates electrons to ubiquinone, which feeds electrons into complex III of the ETC.

Remember that one molecule of NADH and $FADH_2$ generates ~2.5 ATP and ~1.5 ATP, respectively. Therefore, glycolysis generates 2 ATP and 2 NADH (5 ATP); pyruvate oxidation to two molecules of acetyl-CoA generates 2 H (5 ATP) for a total of 12 ATP molecules. Each acetyl-CoA through the TCA cycle generates 3 NADH (7.5 ATP), 1 $FADH_2$ (1.5 per each cycle), and 1 GTP (can be converted to ATP). Thus, one molecule of acetyl-CoA through the TCA cycle generates 10 ATP molecules and 20 ATP molecules from entry of two acetyl-CoA into the TCA cycle. The complete oxidation of glucose by the PDH complex and TCA cycle leads to the production of 30 molecules of ATP and six $CO_2$ molecules as waste.

## MULTIPLE FACTORS CONTROL CELLULAR RESPIRATION

Initial work in the 1950s by Britton Chance and G.R. Williams proposed that the respiratory rate in cells is controlled by cellular ATP use. In this model, the increase in cellular ATP use decreases cytosolic ATP levels and increases cytosolic ADP and $P_i$

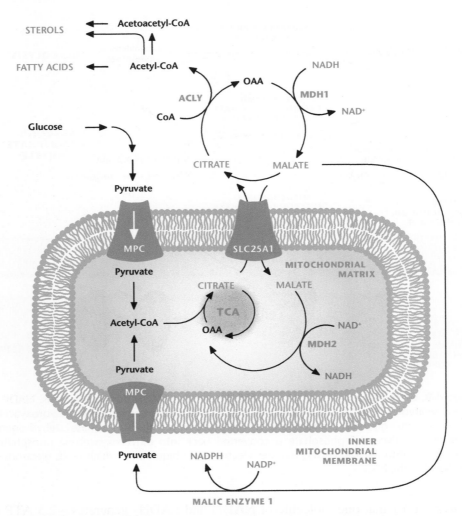

**Figure 4-9.** Citrate transporter. Citrate is transported into the cytosol by the SLC25A1 transporter in exchange for malate. Citrate in the cytosol is converted into acetyl-CoA and oxaloacetate (OAA) by ACLY. Acetyl-CoA is a precursor for sterol and fatty acid production. MDH1 converts the OAA into malate, which is transported into the mitochondrial matrix in exchange for citrate. Malate can also be converted into pyruvate by malic enzyme 1 (ME1).

levels. The increase in cytosolic ADP levels leads to an increase in mitochondrial ADP via the increased activity of the adenine nucleotide translocase (ANT). The increased mitochondrial ADP concentration stimulates complex V to augment the rate of ATP synthesis, which results in a decrease in the mitochondrial membrane potential, thus stimulating the respiratory chain to consume oxygen.

In the ensuing decades, it has become clear that other factors also control the respiratory rate, such as the availability of reducing equivalents provided by the TCA cycle, electron flux through the ETC, the availability of ADP provided by cellular

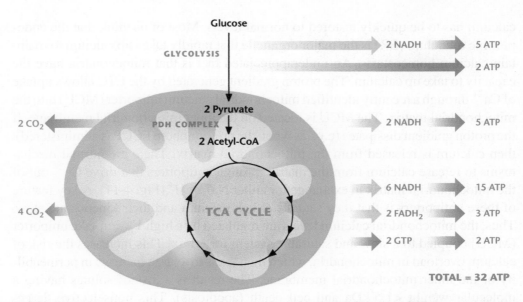

**Figure 4-10.** Glucose oxidation generates 32 ATP. NADH generates 2.5 ATP, whereas $FADH_2$ generates 1.5 ATP. Glucose oxidation through glycolysis and TCA cycle generates 10 NADH (25 ATP) and 2 $FADH_2$ (3 ATP). Glycolysis and TCA cycle generate 2 ATP each through substrate level phosphorylation.

ATPases, ANT, and the magnitude of the proton leak. How to quantitatively measure the relative control exerted by these processes on the respiratory rate remained elusive until the early 1970s, when metabolic control analysis was developed. This helped determine control coefficients of a particular protein over metabolic flux through a pathway. The control coefficient is the percent change in the respiratory rate divided by the percent change in the protein or complex causing the change in the respiratory rate. For example, if a 10% change in the ANT results in a 10% change in the respiratory rate, then the control coefficient of the proton leak would be 1. However, if the 10% change resulted in 1% change in the respiratory rate, then the control coefficient would be 0.1. In the late 1980s, metabolic control analysis on isolated rat hepatocytes determined that 15%–30% of respiration is controlled by the NADH supply (this includes the pyruvate supply to the mitochondria, the TCA cycle, and any other NADH-supplying reaction), 20% is controlled by proton leakage, and 0%–15% is controlled by the ETC. The remaining 50% is controlled by ATP synthesis, transport, and use, of which, in most cells, the dominant factor is the rate of ATP use by cellular processes.

## MITOCHONDRIA REGULATE CALCIUM HOMEOSTASIS

Calcium is an essential ion for multiple processes in the cell, and maintenance of calcium homeostasis is crucial for cells to function properly. An increase in cytosolic calcium has to be quickly restored to normal levels. Most of us think that the endo-plasmic reticulum (ER) is the major organelle that rapidly takes up calcium to maintain calcium homeostasis. An underappreciated fact is that mitochondria have the capacity to take up calcium. The proton gradient generated by the ETC allows uptake of $Ca^{2+}$ through a recently identified mitochondrial calcium uniporter (MCU) into the mitochondrial matrix. The MCU is located on the inner mitochondrial membrane. If the proton gradient dissipates (e.g.. as when the protonophore FCCP is administered), then calcium is released from the mitochondrial matrix. There are normal mechanisms to release calcium from the matrix through antiporters that drive $Ca^{2+}$ out of the mitochondrial matrix in exchange for either $Na^+$ or $H^+$ (Fig. 4-11). A key feature of these antiporters is that they saturate at low calcium and their kinetics are slow. Thus, the mitochondrial calcium levels are regulated by a high $V_{max}$ uptake uniporter (MCU) coupled to a slow and saturable system for efflux. This increases the risk of calcium overload in mitochondria, which then leads to a large increase in permeability of the inner mitochondrial membrane. The result is a loss of solutes having a molecular weight <1.5 kDa and cell death (apoptosis). This nonselective, large-conductance channel is called the mitochondrial permeability transition pore (MPTP). Mitochondrial calcium overload that triggers MPTP formation has been implicated in heart and brain ischemic injury, as well as in several neurodegenerative disorders, including Parkinson's disease. However, calcium uptake in the physiological range stimulates oxidative phosphorylation through allosteric activation of PDH, isocitrate dehydrogenase, and α-ketoglutarate dehydrogenase, as well as stimulation of the ATP synthase (complex V) and ANT. This allows for a coordinated up-regulation of oxidative phosphorylation caused by elevated mitochondrial calcium. Many cells increase their metabolic demand as a result of an increase in calcium. For example, the increase in cytosolic calcium triggers muscle contractility. This increased cytosolic calcium is sequestered in the mitochondrial matrix in which concomitant up-regulation of oxidative phosphorylation provides sufficient ATP to sustain contractility.

## MITOCHONDRIAL BIOSYNTHETIC ACTIVITY CAN BE UNCOUPLED FROM ATP-GENERATING CAPACITY

The earliest TCA cycle evolved in the absence of oxygen, thus it was not an oxidative clockwise cycle but rather a reductive counterclockwise cycle in which NADH or NADPH was consumed to drive the cycle and generate the intermediates that served as precursors for biosynthesis. Can mammalian cells generate TCA-cycle intermediates without generating ATP? There are instances in which the ETC is inhibited when cells are exposed to low-oxygen conditions or proteins of the ETC are mutated in certain diseases. In these scenarios, glycolysis has a remarkable capacity to generate the necessary ATP to sustain cell survival, and this ATP is imported into the matrix to

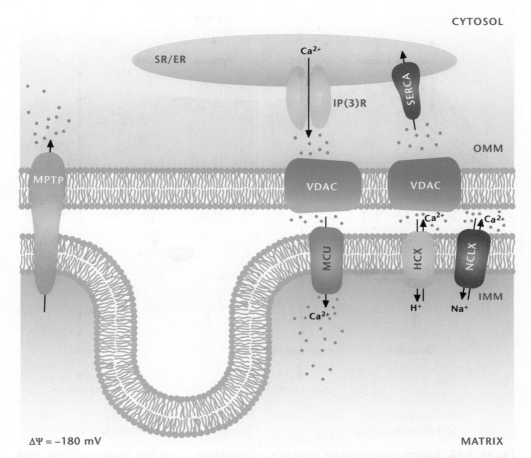

**Figure 4-11.** Mitochondria regulate calcium homeostasis. The major organelle that regulates calcium is the ER, which takes up calcium into the ER by sarco/ER $Ca^{2+}$-ATPase (SERCA) and releases calcium from the ER into the cytosol by inositol trisphosphate receptor (InsP3R). Mitochondria can also regulate calcium by sequestering cytosolic calcium through the MCU. Calcium is transported from mitochondria into the cytosol through the mitochondrial $H^+$–$Ca^{2+}$ exchanger (HCX) and mitochondrial $Na^+$–$Ca2^+$ exchanger (NCLX) transporters. (Adapted from Raffaello et al. 2012, with permission from Elsevier.)

generate a mitochondrial inner membrane potential by complex V hydrolysis of ATP. But, how are TCA-cycle intermediates generated in this case? As we learned earlier in this chapter, citrate is an essential TCA-cycle intermediate because it can be exported through the inner and outer mitochondrial membranes into the cytosol, where it is converted into acetyl-CoA and OAA by the enzyme ACLY. Cells with functional mitochondria generate acetyl-CoA from pyruvate and combine it with OAA to generate citrate, which is exported out to the cytosol, resulting in depletion of TCA-cycle intermediates, including OAA (Fig. 4-12). Glutaminolysis provides α-ketoglutarate, which enters the TCA cycle to generate OAA, which, in turn, condenses with pyruvate-generated acetyl-CoA to produce citrate.

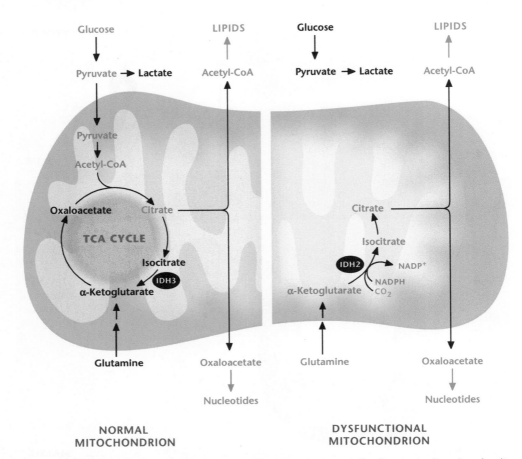

**Figure 4-12.** Glutamine-dependent reductive carboxylation. Cells that have functional mito-chondria can use glutamine to generate α-ketoglutarate to produce OAA and use pyruvate to gen-erate acetyl-CoA. OAA and acetyl-CoA generate citrate, which can be exported into the cytosol to produce de novo lipids and nucleotides. Cells that have dysfunctional mitochondria due to loss-of-function mutations of proteins in the TCA cycle after the α-ketoglutarate dehydrogenase step or in the ETC cannot oxidize pyruvate into acetyl-CoA. They convert glutamine into α-ketoglutarate, which subsequently becomes citrate through a reverse IDH2-dependent reaction.

In mammalian cells with dysfunctional mitochondria, the TCA cycle can partly reverse itself to generate citrate through a process called glutamine-dependent reductive carboxylation. This recently discovered mechanism begins with conversion of glutamine to glutamate, which is converted to α-ketoglutarate and enters the TCA cycle. The five-carbon α-ketoglutarate is converted by two subsequent reactions catalyzed by NADPH-dependent isocitrate dehydrogenase 2 (IDH2) and aconitase to generate six-carbon citrate molecules through a carboxylation reaction using the reductive power of NADPH (Fig. 4-12). This finding illustrates that modern mitochondria also display robust metabolic plasticity just as their bacterial evolutionary predecessors have.

## MITOCHONDRIA ARE SIGNALING ORGANELLES

The two major roles of mitochondria—the production of energy and support of biosynthesis—make them central to diverse biological outcomes, including proliferation, differentiation, and adaptation to stress. The classical conception of biological outcomes is that they are driven by commands from the nucleus, and changes in mitochondrial metabolism occur simply as a consequence of these commands. Mitochondria are rarely considered to dictate commands or provide signals themselves to change biological outcomes. However, should the cell commit to a process like proliferation or differentiation without adequate functioning mitochondria, it would likely undergo a metabolic crisis resulting in cell death or senescence. For optimal cell function, a health-status-feedback mechanism should exist in mitochondria to act as a checkpoint before cellular action. This feedback is analogous to the fuel gauge on your car, which predicts the distance you may drive. Thus, a mitochondrial checkpoint is required before cells commit to distinct biological outcomes, such as proliferation or differentiation. This reasoning suggests that mitochondria play a causal role in determining biological outcomes.

Over the past 20 years, there have been many studies showing that mitochondria play a vital role in signaling by releasing (1) metabolites, such as acetyl-CoA for protein acetylation, (2) proteins, such as cytochrome $c$, that induce caspase-dependent cell death, and (3) ROS, which can either inhibit or activate certain signaling pathways. Changes in mitochondrial ATP generation can also be transmitted to cytosol through the activation of AMP-activated protein kinase (AMPK). This kinase responds to a decrease in ATP levels concomitant with an increase in AMP levels. AMPK causes cells to decrease anabolic functions and go into a catabolic state to generate energy.

Mitochondria also control signaling by serving as a scaffold for signaling complexes through the tethering of signaling proteins to the outer mitochondrial membrane. One example is mitochondrial antiviral-signaling protein, which is important for appropriate responses to viral infection. Why this protein has to be on the outer mitochondrial membrane remains a mystery. But, there is a growing number of proteins that participate in signaling pathways that are in close proximity to mitochondria.

An exciting new area in mitochondrial signaling is the field of mitochondrial dynamics. Mitochondria are constantly undergoing fission and fusion, as well as moving around the cell. Depending on the environmental or internal signals, mitochondria can move around the cell to appropriate sites to participate in signaling, as well as undergo fission or fusion. Typically, fission-produced mitochondria tend to be less robust bioenergetically than fused mitochondria.

Finally, cells that have undergone necrosis and spilled their contents into the blood can invoke inflammation. Recent studies indicate that mtDNA can activate Toll-like receptor signaling on monocytes, resulting in activation of inflammatory cytokines. Going forward, one of the most exciting aspects of mitochondrial biology will be to understand how mitochondria participate in regulating signaling events in the cytosol to dictate biological outcomes.

## MITOCHONDRIA AND DISEASE

One of the most devastating classes of diseases are those linked to mutations that impair oxidative phosphorylation; these affect at least 1 in 5000 live births. Mitochondrial diseases are caused by either inherited or spontaneous mutations in mitochondrial or nuclear DNA that change the function of mitochondrial proteins or RNA molecules. As mentioned earlier, mtDNA encodes for only 37 of the 3000 genes that create a mitochondrion. Many of the mtDNA-encoded genes make proteins that are critical subunits of the ETC. Several of these mitochondrial diseases occur early in infancy. They present with lactic acidosis, blindness, peripheral neuropathy, skeletal myopathy, deafness, and neurodegeneration, and cause tremendous suffering in the affected patient. Many of these diseases cause extensive damage to cells of the skeletal muscles, brain, heart, and liver.

One common early-onset mitochondrial disorder is MELAS (mitochondrial encephalomyopahty, lactic acidosis, and stroke-like episodes). MELAS is mainly caused by mutations in genes encoded by mDNA for mitochondrial-specific transfer RNAs, and it is maternally inherited. Interestingly, different mutations in mitochondrial and nuclear DNA can result in the same diseases. An example is Leigh syndrome, which is characterized by abnormalities in the brain stem, cerebellum, and basal ganglia and often displays elevated lactic acidosis. Mitochondrial disease symptoms can also by mimicked by drugs that have off-target toxicity. Fialuridine, an antiviral drug for hepatitis B, caused lactic acidosis, neuropathy, hepatic failure, and myopathy in a subset of patients in a phase II clinical trial because of an off-target effect on mitochondrial function, and the trial was halted. Interestingly, many widely prescribed therapeutics, including statins, antibiotics, and the HIV drug, zidovudine, are known to have off-target effects on mitochondria that impair function.

An exciting area of growth in the mitochondrial-linked diseases is the accumulating evidence that mitochondrial dysfunction is connected to several common diseases, such as diabetes, obesity, cancer, and neurodegenerative disorders. Progressive decline in the expression of mitochondrial genes is also a prominent feature of normal human aging, whereas caloric restriction, which may increase fitness and life span in humans, causes a robust increase in mitochondrial biogenesis. The big unanswered question is whether the decline in mitochondrial function during normal aging or many of the common diseases is causal or a consequence. In fact, it is not fully understood whether slight mitochondrial impairment observed in some diseases or in aging is beneficial or harmful. Recent data in model organisms ranging from yeast to mice indicate that diminishing mitochondrial function can promote metabolic health and longevity.

## REFERENCES

Anso E, Mullen AR, Felsher DW, Matés JM, Deberardinis RJ, Chandel NS. 2013. Metabolic changes in cancer cells upon suppression of MYC. *Cancer Metab* **1:** 7.

Raffaello A, De Stefani D, Rizzuto R. 2012. The mitochondrial $Ca^{2+}$ uniporter. *Cell Calcium* **52:** 16–21.

## ADDITIONAL READING

Brand MD, Kesseler A. 1995. Control analysis of energy metabolism in mitochondria. *Biochem Soc Trans* **23:** 371–376.

Brand MD, Nicholls DG. 2011. Assessing mitochondrial dysfunction in cells. *Biochem J* **435:** 297–312.

Chance B, Williams GR. 1955. Respiratory enzymes in oxidative phosphorylation. III. The steady state. *J Biol Chem* **217:** 409–427.

Chandel NS. 2014. Mitochondria as signaling organelles. *BMC Biol* **12:** 34.

De Stefani D, Rizzuto R. 2014. Molecular control of mitochondrial calcium uptake. *Biochem Biophys Res Commun* **449:** 373–376.

Gray MW. 2012. Mitochondrial evolution. *Cold Spring Harb Perspect Biol* **4:** a011403.

Krebs HA. 1937. The intermediate metabolism of carbohydrates. *Lancet* **230:** 736–738.

Krebs HA, Johnson WA. 1937. The role of citric acid in intermediate metabolism in animal tissues. *Enzymologia* **4:** 148–156.

Lane N. 2014. Bioenergetic constraints on the evolution of complex life. *Cold Spring Harb Perspect Biol* **6:** a015982.

Lapuente-Brun E, Moreno-Loshuertos R, Acín-Pérez R, Latorre-Pellicer A, Colás C, Balsa E, Perales-Clemente E, Quirós PM, Calvo E, Rodríguez-Hernández MA, et al. 2013. Supercomplex assembly determines electron flux in the mitochondrial electron transport chain. *Science* **340:** 1567–1570.

Lindley D, Clarke M. 1988. Nobel prizes announced for physics and for chemistry. *Nature* **335:** 752–753.

Liu X, Kim CN, Yang J, Jemmerson R, Wang X. 1996. Induction of apoptotic program in cell-free extracts: Requirement for dATP and cytochrome *c*. *Cell* **86:** 147–157.

Martin W, Müller M. 1998. The hydrogen hypothesis for the first eukaryote. *Nature* **392:** 37–41.

McKenzie R, Fried MW, Sallie R, Conjeevaram H, Di Bisceglie AM, Park Y, Savarese B, Kleiner D, Tsokos M, Luciano C, et al. 1995. Hepatic failure and lactic acidosis due to fialuridine (FIAU), an investigational nucleoside analogue for chronic hepatitis B. *N Engl J Med* **333:** 1099–1105.

Mitchell P. 1961. Coupling of phosphorylation to electron and hydrogen transfer by a chemi-osmotic type of mechanism. *Nature* **191:** 144–148.

Murphy MP. 2008. How mitochondria produce reactive oxygen species. *Biochem J* **417:** 1–13.

Nunnari J, Suomalainen A. 2012. Mitochondria: In sickness and in health. *Cell* **148:** 1145–1159.

Pagliarini DJ, Rutter J. 2013. Hallmarks of a new era in mitochondrial biochemistry. *Genes Dev* **27:** 2615–2627.

Sena LA, Chandel NS. 2012. Physiological roles of mitochondrial reactive oxygen species. *Mol Cell* **48:** 158–167.

Vafai SB, Mootha VK. 2012. Mitochondrial disorders as windows into an ancient organelle. *Nature* **491:** 374–383.

Weber BH. 1991. Glynn and the conceptual development of the chemiosmotic theory: A retrospective and prospective view. *Biosci Rep* **11:** 577–617.

West AP, Shadel GS, Ghosh S. 2011. Mitochondria in innate immune responses. *Nat Rev Immunol* **11:** 389–402.

# 5

## NADPH—The Forgotten Reducing Equivalent

I N CHAPTER 4, THE REDUCING EQUIVALENTS NADH and $FADH_2$ were essential products of the TCA that were used by the electron transport chain (ETC) to generate ATP through oxidative phosphorylation. But an often-overlooked reducing equivalent is NADPH, which is not used for generating ATP but for biosynthesis of macromolecules. NADPH is the major reducing equivalent driving de novo synthesis of fatty acids, cholesterol, amino acids, and nucleotides. Its other major function is generation of superoxide ($O_2^-$) by NADPH oxidases (NOXs) and scavenging of $H_2O_2$ by regenerating GSH and the antioxidant protein thioredoxin (TRX). Its role in generating superoxide and scavenging $H_2O_2$ will be covered in this chapter, and the biosynthetic reactions coupled to NADPH will be covered over the next few chapters.

NADPH can be considered a high-energy molecule similar to NADH. However, the electrons of NADPH are used for biosynthesis of macromolecules and the scavenging and generation of ROS, whereas the electrons of NADH are ultimately transferred by the ETC to oxygen. Cells maintain a high-NADPH/$NADP^+$ ratio. Many biosynthetic reactions are coupled to $NADP^+ + 2e^- + 1H^+ \rightarrow NADPH$ to make these reactions thermodynamically favorable. Accordingly, cells have multiple sources of NADPH that are generated by cytosolic glycolytic and mitochondrial TCA-cycle intermediates. Conceptually, this makes sense because as these intermediates are used as precursors for biosynthesis of macromolecules, they concomitantly generate NADPH to drive the biosynthetic reactions thermodynamically.

### QUICK GUIDE TO NADPH (FIG. 5-1)

- The glycolytic intermediate glucose 6-phosphate can funnel into the pentose phosphate pathway (PPP) to generate NADPH in the cytosol.

- Isocitrate dehydrogenase 1 (IDH1) and 2 (IDH2) generate NADPH in the cytosol and mitochondrial matrix, respectively.

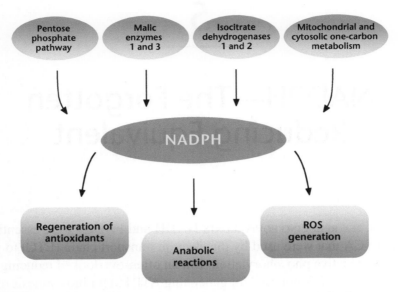

**Figure 5-1.** Diverse roles of NADPH in metabolism. There are multiple sources that generate NADPH in the mitochondria and cytosol. NADPH is critical for many anabolic reactions and is essential to maintain antioxidant capacity in cells. NADPH can also be used to generate ROS through NADPH oxidases.

- Malic enzymes 1 (ME1) and 3 (ME3) generate NADPH in the cytosol and mitochondrial matrix, respectively.

- One-carbon metabolism generates NADPH in the cytosol and mitochondrial matrix.

- NADPH is used for reductive biosynthesis of fatty acids, cholesterol, nucleotides, and amino acids.

- NADPH is used by NOXs to generate superoxide.

- NADPH is used as an electron donor to reduce oxidized GSH and TRX, thus maintaining antioxidant capacity in cells.

## GLYCOLYTIC INTERMEDIATES GENERATE CYTOSOLIC NADPH THROUGH THE PENTOSE PHOSPHATE PATHWAY (PPP)

The oxidative phase of the pentose phosphate pathway (PPP), also called the phosphogluconate pathway and hexose monophosphate shunt, is one of the primary sources of cytosolic NADPH (Fig. 5-2). This anabolic pathway is initiated when glucose is converted into glucose 6-phosphate, which can either go down the glycolysis pathway or enter the PPP to generate NADPH. The end product of the oxidative phase of the PPP is ribulose 5-phosphate, which enters the nonoxidative phase of the PPP to

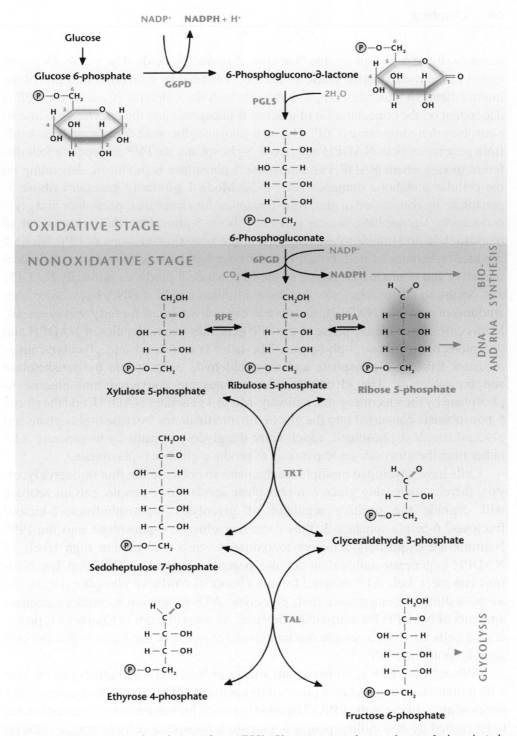

**Figure 5-2.** The pentose phosphate pathway (PPP). Glucose conversion to glucose 6-phosphate by G6PD in step 1 of glycolysis allows for the funneling of the latter metabolite down the glycolytic pathway or into the PPP. The oxidative stage of the PPP generates NADPH, whereas the nonoxidative stage generates ribose 5-phosphate, which can be used for DNA and RNA synthesis. G6PD, glucose 6-phosphate hydrogenase; PGLS, 6-phosphogluconolactonase; 6PGD, 6-phosphogluconate dehydrogenase; RPE, ribulose 5-phosphate-3-epimerase; RPIA, ribose 5-phosphate isomerase A; TKT, transketolase; TAL, transaldolase.

generate ribose 5-phosphate, the "backbone" of nucleic acids. The metabolites generated in the nonoxidative phase of the PPP can also be converted into glycolytic intermediates to generate ATP. The flux through the oxidative phase of the PPP is dependent on the concentration of glucose 6-phosphate and the enzymes glucose 6-phosphate dehydrogenase (G6PDH) and 6-phosphogluconate dehydrogenase. Aside from generating both NADPH and ribose 5-phosphate, the PPP can operate four different modes where NADPH and/or ribose 5-phosphate is produced, depending on the cellular metabolic demands (Fig. 5-3). Mode 1 primarily generates ribose 5-phosphate by conversion of glucose 6-phosphate into fructose 6-phosphate and glyceraldehyde 3-phosphate, which generate ribose 5-phosphate through reversal of transaldolase and transketolase reactions in the nonoxidative stage of PPP. Mode 2 produces a balance of both NADPH and ribose 5-phosphate through the canonical oxidative and nonoxidative stages of the PPP. Mode 3 produces primarily NADPH and occurs in postmitotic cells that have a limited need for DNA but require high amounts of cytosolic NADPH, such as adipose or liver tissue, for fatty acid synthesis. In this case, the oxidative phase of the PPP generates two molecules of NADPH and one molecule of ribose 5-phosphate. The latter is converted into glycolytic intermediates fructose 6-phosphate and glyceraldehyde 3-phosphate by transketolase and transaldolase. The glycolytic intermediates are converted into glucose 6-phosphate by the gluconeogenic pathway. Mode 4 generates NADPH, and the ribose 5-phosphate is converted into the glycolytic intermediates fructose 6-phosphate and glyceraldehyde 3-phosphate, which enter the glycolytic pathway to generate ATP rather than the gluconeogenic pathway to produce glucose 6-phosphate.

Cells have developed multiple mechanisms to decrease the flux through glycolysis, thereby increasing glucose 6-phosphate levels. For example, certain neurons will degrade the positive regulator of glycolysis 6-phosphofructo-2-kinase/fructose-2,6-bisphosphatase-3, thus funneling glucose 6-phosphate into the PPP. Neurons are exquisitely sensitive to oxidative stress and require high levels of NADPH to generate antioxidant defense systems regulated by GSH and Trx. Neurons can meet their ATP demand by mitochondrial oxidative phosphorylation and so are willing to compromise their glycolytic ATP generation to produce copious amounts of NADPH for antioxidant defense. As we will learn in Chapter 11, proliferating cells also have multiple mechanisms to ensure they can balance flux through glycolysis and the PPP.

Although the PPP is an important source of NADPH, individuals can survive with mutations in the oxidative phase of this pathway. (The next section covers other major sources of cytosolic PPP.) The most common human enzyme defect estimated to be carried by 400 million people is glucose 6-phosphate dehydrogenase (G6PD) deficiency (see Box 5-1). The oxidative phase of the PPP is the only source of NADPH in red blood cells, which are devoid of mitochondria. As a result, these cells become very sensitive to oxidative stress, and anemia can develop because of damaged red blood cells.

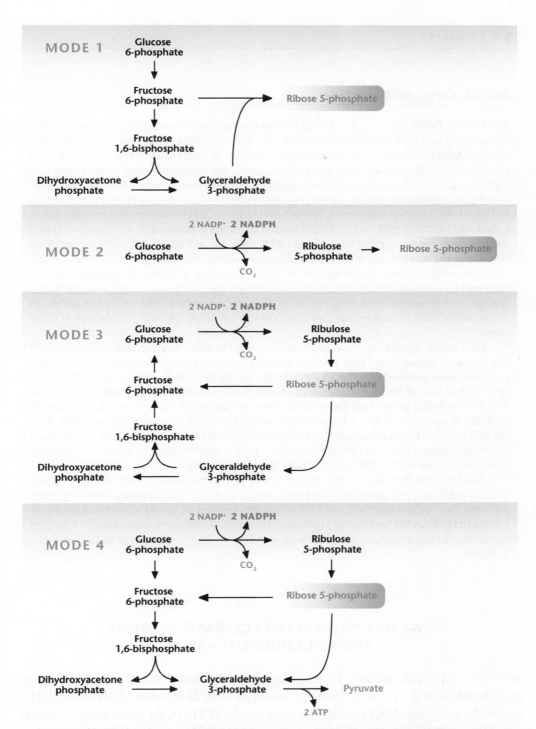

**Figure 5-3.** Different modes of the pentose phosphate pathway (PPP). The PPP can operate in different modes depending on cellular requirements of NADPH and ribose 5-phosphate. Mode 1 primarily generates ribose 5-phosphate by funneling glycolytic intermediates into the nonoxidative stage of PPP. Mode 2 produces balance of both NADPH and ribose 5-phosphate. Mode 3 produces primarily NADPH as the ribose 5-phosphate is funneled into glycolysis to go back into the oxidative stage of PPP for further generation of NADPH. Mode 4 generates NADPH and the ribose 5-phosphate is funneled into glycolysis to generate ATP. (Adapted, with permission, from Berg et al. 2012, © W.H. Freeman and Company.)

---

**BOX 5-1. G6PD—NOW AND THEN**

---

The X-linked G6PD deficiency is the most common genetic enzyme disorder in humans, affecting an estimated 400 million people worldwide. Its distribution in northern Africa, southern Asia, the Mediterranean countries, and Middle Eastern regions matches that of high-malaria (*Plasmodium falciparum*–caused) prevalence. G6PD deficiency provides some protection against malaria, as do the thalassemias and sickle cell disease also common in these areas, suggesting an evolutionary benefit. G6PD converts $NADP^+$ to NADPH in the PPP and this protects red blood cells from oxidative stress. The lack of NADPH in G6PD deficiency results in the failure to reduce $H_2O_2$ and ROS, oxidation of hemoglobin to methemoglobin, and the membrane damage that results in hemolysis. These G6PD-deficient red blood cells, when parasitized by the malaria parasite, are more readily phagocytosed by macrophages.

The most common manifestation of G6PD deficiency is hemolytic anemia, which is triggered by infection and certain drugs. But, perhaps the strangest is favism, a severe hemolytic reaction to eating or inhaling the pollen of the broad or fava bean, *Vicia faba*. Although favism occurs only in people with G6PD deficiency, not all people with this deficiency suffer favism.

Fava beans, apparently, have been known since Neolithic times, but stories of avoiding the ingestion of fava beans date back to ancient Greece and the Pythagoreans. The reasons vary, but many believed that these beans contained the souls of the dead and that the black spot on the hilum of the bean was associated with death. Aristotle tells of Pythagoras (ca. 570–495 BCE) refusing to eat fava beans, and others tell stories of Pythagoras refusing to walk through fields of fava beans. Although there are no contemporary accounts, later historians tell of Pythagoras being pursued by his enemies and refusing to cross a bean field. As described by the 3rd century historian Diogenes Laertius, "sooner than trample on it [i.e., a bean field], he endured to be slain at the cross-roads by the men of Acragas."

Modern reports of favism date from the mid-19th century when some G6PD-deficient patients presented with jaundice. Fava beans contain glycosidic compounds, including vicine and covicine, that in the presence of G6PD deficiency result in increased susceptibility of erythrocytes to $H_2O_2$ and other ROS, and ultimately hemolysis. G6PD is the major source of NADPH in red blood cells; thus lack of this enzyme makes them highly susceptible to ROS-induced damage.

## TCA-CYCLE INTERMEDIATES GENERATE CYTOSOLIC AND MITOCHONDRIAL NADPH

Isocitrate and malate are two TCA-cycle metabolites that generate NADPH through mitochondrial and cytosolic forms of isocitrate dehydrogenases (IDH1 and IDH2) and malic enzymes (ME1 and ME3), respectively. IDH2 in the mitochondrial matrix converts isocitrate to α-ketoglutarate to generate NADPH from $NADP^+$. Note that isocitrate dehydrogenase 3 (IDH3) converts isocitrate to α-ketoglutarate to generate NADH from $NAD^+$. IDH1 in the cytoplasm converts isocitrate to α-ketoglutarate to generate NADPH from $NADP^+$. Cytoplasmic pools of isocitrate are generated from citrate by cytosolic aconitase 1. The generation of NADPH by IDH1 requires citrate export from the mitochondria (Fig. 5-4) in exchange for malate imported into the mitochondria.

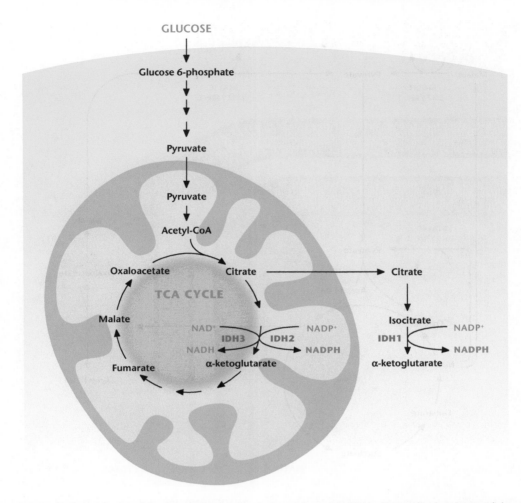

**Figure 5-4.** Isocitrate dehydrogenases 1 and 2 produce NAPDH. Cytosolic isocitrate dehydrogenase 1 (IDH1) and mitochondrial isocitrate dehydrogenase 2 (IDH2) produce NADPH through conversion of isocitrate into α-ketoglutarate. Note that mitochondrial isocitrate dehydrogenase 3 (IDH3) generates NADH by converting isocitrate into α-ketoglutarate.

The mitochondrial malate is used for exchange of cytoplasmic α-ketoglutarate; therefore, this cycle produces no net change in the cytoplasmic malate pool.

The citrate pool in the cytoplasm can also be converted into oxaloacetate and acetyl-CoA by ATP-citrate lyase (ACLY). The acetyl-CoA can be used for protein acetylation or fatty acid synthesis. Malate dehydrogenase converts the oxaloacetate into malate, which is subsequently converted into pyruvate by ME1 to generate NADPH (Fig. 5-5). The pyruvate generated is transported into the mitochondria to enter the TCA cycle to generate citrate. ME2 can generate NADPH by converting mitochondrial matrix pool of malate into pyruvate that can reenter the TCA cycle.

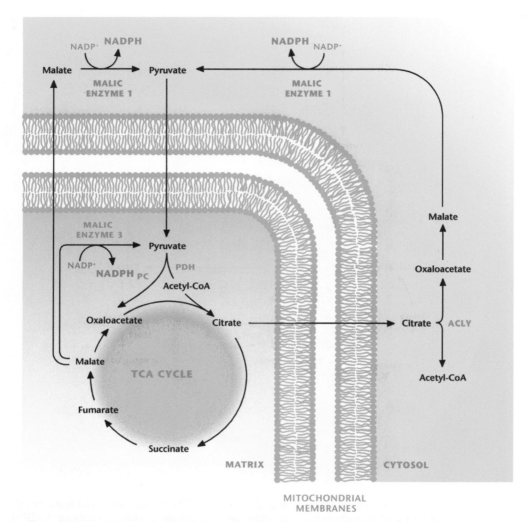

**Figure 5-5.** Malic enzymes 1 and 3 produce NADPH. Mitochondrial malic enzyme 3 can generate NADPH by converting malate into pyruvate. Malate can also exit from the TCA cycle where it is converted into pyruvate by malic enzyme 1, resulting in the production of NADPH. The TCA-cycle intermediate citrate can be transported into the cytosol where it is converted into acetyl-CoA and oxaloacetate by ATP-citrate lyase (ACLY). The latter is converted into malate to produce NADPH by malic enzyme 1. PC, pyruvate carboxylase; PDH, pyruvate dehydrogenase.

## CYTOSOLIC AND MITOCHONDRIAL ONE-CARBON METABOLISM GENERATES NADPH

One-carbon metabolism integrates the amino acids serine and glycine into the folate and methionine cycles that participate in many diverse biological outputs, including nucleotide synthesis (see Chapter 9) and methylation reactions involved in epigenetics (see Chapter 10). In this metabolism, a carbon unit from serine or glycine is transferred to tetrahydrofolate (THF) to form 5,10-methylene-THF. Recent studies have highlighted the up-regulation and necessity of one-carbon metabolism for tumor cell proliferation, thus making this metabolic pathway a viable therapeutic target (see Chapter 11). One-carbon metabolism comprises parallel cytosolic and mitochondrial pathways (Fig. 5-6). Although there is restricted exchange between the cytoplasmic and mitochondrial pools of THF molecules, these compartments are metabolically connected by transport of the one-carbon donors (serine, glycine, and formate) across the mitochondrial membranes. Recent studies indicate that the one-carbon metabolism is a significant source of NADPH in the mitochondria and, to a lesser extent, in the cytoplasm. These NADPH-generating reactions are

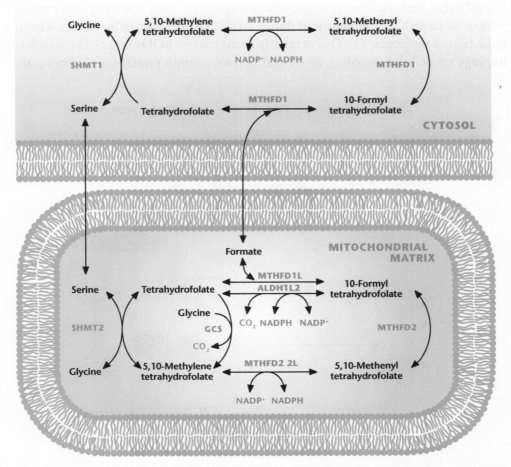

**Figure 5-6.** Mitochondrial and cytosolic one-carbon metabolism generates NADPH. The one-carbon amino acids serine and glycine can feed into folate cycle to generate NADPH. The cytosolic enzyme MTHFD1 and mitochondrial matrix enzymes MTHFD1L, MTHFD2L, and ALDH1L2 generate NADPH. One-carbon metabolism is a significant source of NADPH in the mitochondria. GCS, glycine cleavage system.

catalyzed by MTHFD1 in the cytosol and MTHFD1L, MTHFD2L, and ALDH1L2 in the mitochondrial matrix compartment (Fig. 5-6).

## THIOL-DEPENDENT REDOX SIGNALING

Subsequent chapters will cover the essential role of NADPH in multiple anabolic reactions to generate lipids, nucleotides, and amino acids. As mentioned earlier, a major role of NADPH is to detoxify ROS, which are intracellular chemical species that contain oxygen ($O_2$) and are reactive toward lipids, proteins, and DNA. ROS include $O_2^-$, $H_2O_2$, and hydroxyl radicals ($OH^-$) (Fig. 5-7). ROS are more chemically reactive than $O_2$ and are able to trigger various biological events. Each ROS has different intrinsic chemical properties that dictate their reactivity and preferred biological targets. $O_2^-$ is produced during oxidative metabolism by the one-electron reduction of molecular $O_2$. $O_2^-$ is rapidly converted by SODs into $H_2O_2$, which can impinge on cellular signaling by oxidizing thiols within proteins and altering their

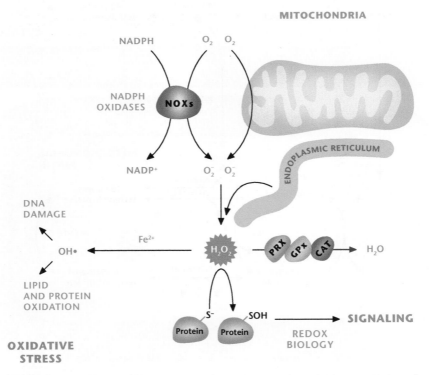

**Figure 5-7.** Sources of ROS. Mitochondria, NADPH oxidase (NOXs), and the endoplasmic reticulum are three major sources of $H_2O_2$, which can either be detoxified into $H_2O$ by peroxiredoxins (PRXs), catalase (CAT), or glutathione peroxidases (GPXs) or activate signaling pathways by oxidizing specific thiols within proteins. $H_2O_2$ in the presence of iron ($Fe^{2+}$) can generate hydroxyl radical ($OH^-$) that incurs damage to DNA, lipids, and proteins.

function (Fig. 5-8). The concentration of $H_2O_2$ associated with signaling is likely in the low-nanomolar range. Unlike $O_2^-$, $H_2O_2$ can readily diffuse through membranes, making it an ideal intracellular signaling molecule. ROS have been traditionally thought of as toxic metabolic byproducts that cause cellular damage. However, studies throughout the past decade have highlighted the role of $H_2O_2$ in cell signaling. $H_2O_2$ is detoxified to water by glutathione peroxidases (GPXs), peroxiredoxins (PRXs), and catalase. In the presence of ferrous or cuprous ions, $H_2O_2$ can become a hydroxyl radical, which is very reactive and causes oxidation of lipids, proteins, and DNA, resulting in damage to the cell.

ROS levels associated with signaling comprise redox biology, whereas ROS levels associated with pathology are referred to as oxidative damage. ROS have been

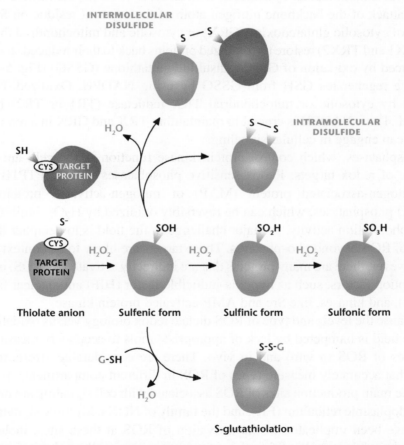

**Figure 5-8.** $H_2O_2$-dependent signaling. $H_2O_2$ regulates signaling pathways by oxidation of thiol groups on cysteines within proteins that show a low $pK_a$, allowing the cysteine thiol group to exist as a thiolate anion ($S^-$). $H_2O_2$ readily oxidizes thiolate, yielding $SO^-$. Under high concentrations of $H_2O_2$, $SO^-$ can undergo further oxidation to generate $SO_2^-$ and $SO_3^-$. $SO^-$ can undergo further modifications including intra- or intermolecular disulfide bonds and S-gluthathiolation. (Modified from Finkel 2011.)

shown to play a causal role in various cellular events, including proliferation, differentiation, metabolic adaptation, and the regulation of adaptive and innate immunity. ROS control cellular signaling by causing reversible posttranslational modifications to proteins that regulate signaling pathways. Biological redox reactions catalyzed by $H_2O_2$ or organic hydroperoxide typically involve the oxidation of thiol groups on cysteines with a low $pK_a$, allowing the cysteine thiol group (SH) to exist as a thiolate anion ($S^-$). $H_2O_2$ readily oxidizes thiolate, yielding sulfenic acid ($SO^-$). Under high concentrations of $H_2O_2$, $SO^-$ can undergo further oxidation to generate sulfinic ($SO_2^-$) and sulfonic ($SO_3^-$) acids. $SO_3^-$, generally, are irreversible oxidative modifications (Fig. 5-8). A common mechanism for preventing irreversible oxidation of catalytic cysteines is to incorporate the $SO^-$ intermediate into a disulfide (S–S) bond or sulfenic–amide (S–N) bond. Disulfides are formed by reaction of $SO^-$ with either an inter- or intramolecular cysteine, or with GSH. S–N bonds are formed by nucleophilic attack of the backbone nitrogen atom of the adjacent residue on $SO^-$. The actions of cytosolic glutaredoxins (GRX) or cytosolic and mitochondrial thioredoxins (TRX1 and TRX2) restore the oxidized proteins back to their reduced state. GRXs are reduced by oxidation of GSH to disulfide glutathione (GSSG) (Fig. 5-8). GSH reductase regenerates GSH from GSSG by using NADPH. Oxidized TRXs are reduced by cytosolic or mitochondrial TRX reductase (TR1 or TR2) by using NADPH. Thus, NADPH is crucial in maintaining TRX and GRX in a reduced state so they can engage in cellular signaling.

Phosphatases, which control protein kinase function in the cell, are the best example of redox targets. Redox-sensitive phosphatases include PTP1B, PTEN, and mitogen-associated protein (MAP) or mitogen-activated protein kinase (MAPK) phosphatases, which can be reversibly oxidized by $H_2O_2$, inhibiting their dephosphorylation activity. A major challenge in the field is to decipher the direct targets of ROS beyond phosphatases. These targets are likely to be context dependent. However, there are many proteins that are indirectly activated by ROS, including transcription factors, such as hypoxia-inducible factor (HIF) and nuclear factor-κB (NF-κB), and kinases, like Src and AMP-activated protein kinase.

Because the levels and type of ROS dictate redox biology versus oxidative damage, the field is hampered by lack of appropriate tools to accurately measure levels and types of ROS in vitro and in vivo. There are considerable efforts to design probes that accurately measure types of ROS in different compartments of the cell. The three main production sites of ROS associated with cell signaling are mitochondria, endoplasmic reticulum (ER), and the family of NOXs. Multiple signaling pathways have been implicated in the activation of ROS at these sites, including an increase in calcium, gain of function of oncogenes, as well as changes in oxygen and nutrient levels.

Mitochondria are major generators of ROS at eight distinct sites, as discussed previously in Chapter 4. The three best-characterized sites within the mitochondrial respiratory chain are complexes I, II, and III, located in the inner mitochondrial membrane. These complexes generate $O_2^-$ by the one-electron reduction of molecular $O_2$. Complexes I, II, and III release $O_2^-$ into the mitochondrial matrix, where SOD2 rapidly converts it into $H_2O_2$. Complex III can also generate $O_2^-$ and release it into the intermembrane space where the $O_2^-$ traverses through voltage-dependent anion channels (VDACs) into the cytosol and is converted into $H_2O_2$ by SOD1. Complexes I and III are considered major sites of $O_2^-$ generation (Fig. 5-9). Other sites within mitochondria known to generate $O_2^-$ that are gaining biological significance include pyruvate dehydrogenase, α-ketoglutarte dehydrogenase, glycerol 3-phosphate dehydrogenase, and proline dehydrogenase. An important function of mitochondrial-generated $O_2^-$ and $H_2O_2$ is the activation of cellular signaling, which is essential for stem-cell differentiation, tissue morphogenesis, lymphocyte and

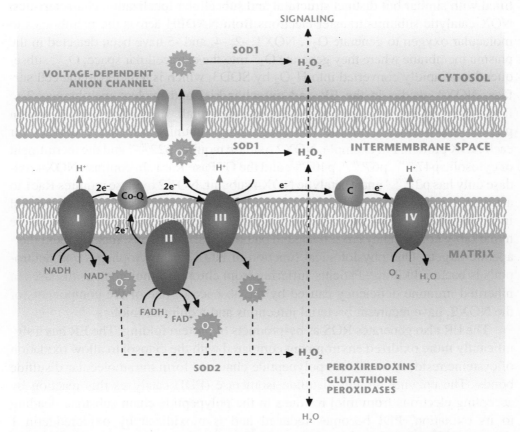

**Figure 5-9.** The mitochondrial electron transport chain (ETC) generates ROS. ETC complexes I, II, and III generate $O_2^-$ superoxide in the mitochondrial matrix that is rapidly converted into $H_2O_2$ by SOD2. $H_2O_2$ can traverse through the inner and outer mitochondrial membranes to activate cellular signaling. Complex III can also generate $O_2^-$ superoxide into the intermembrane space where it can traverse through voltage-dependent anion channels (VDACs) into the cytosol or into $H_2O_2$ by SOD1 in the intermembrane space. Subsequently $H_2O_2$ can efflux into the cytosol to activate signaling.

macrophage activation, and metabolic adaptation under nutrient deprivation. The molecular details of how mitochondrial ROS regulate these diverse biological outcomes are currently being intensely investigated.

The NOX family of proteins is primarily localized to the plasma membrane; however, these proteins can be found on other membranes, including those of the ER and mitochondria. NOXs have also been implicated in cellular signaling and regulation of multiple diverse biological outcomes, including cell proliferation, cell migration, immune cell activation, and metabolic adaptation. NADPH donates electrons to the center of the NOX catalytic subunit to generate $O_2^-$ through the one-electron reduction of $O_2$ (Fig. 5-10). SOD1 in the cytosol converts NOX-generated $O_2^-$ to $H_2O_2$. Seven membrane-bound NOX catalytic isoforms, referred to as NOX1 to NOX5, dual oxidase 1 (DUOX1), and 2 (DUOX2) have been identified with similar but distinct structural and subcellular localization characteristics. NOX catalytic subunits transfer electrons from NADPH across the membranes to molecular oxygen to generate $O_2^-$. NOX1, -2, -4, and -5 have been detected in the plasma membrane where they generate $O_2^-$ into the extracellular space; $O_2^-$, subsequently, is rapidly converted into $H_2O_2$ by SOD3, which is tethered to the cell surface. NOX4 resides in the ER and mitochondrial and nuclear membranes. The facilitation of electron transport requires the small membrane-bound protein $p22^{phox}$ and various cytosolic NOX-regulatory subunits, which are distinct for each NOX protein. For example, NOX2 oxidase requires $p22^{phox}$ and the recruitment of cytosolic $p47^{phox}$, $p67^{phox}$, $p40^{phox}$, and the GTPase Rac1. In contrast, NOX4 oxidase only has $p22^{phox}$ and a catalytic NOX4 subunit. NOX2 oxidase requires Rac1 to activate it optimally, whereas NOX4 oxidase is activated by the transcriptional induction of the NOX4 subunit. Although the NOX family member knockouts are remarkably normal, they are activated in pathological conditions, making them therapeutic targets. One physiological function of NOX2 in macrophages and neutrophils is bacterial killing. Patients suffering from chronic granulomatous disease, an inherited immune deficiency caused by the absence of one of the components of the NOX2, have recurrent bacterial infections and require antibiotics.

The ER also generates ROS as by-products of protein folding. The ER has a significantly more oxidized environment compared with the cytosol to allow oxidation of cysteine residues in nascent polypeptide chains to form intramolecular disulfide bonds. The enzyme protein disulfide isomerase (PDI) catalyzes this reaction by accepting electrons from thiol residues in the polypeptide chain substrate, leading to its oxidation. PDI becomes reduced and is reoxidized by oxidoreductin 1 (ERO1) (Fig. 5-11). FAD accepts two electrons from reduced ERO1 to reoxidize and recycle ERO1 for another round of protein disulfide formation. Subsequently, $FADH_2$ donates its two electrons to $O_2$ resulting in regeneration of FAD and $H_2O_2$ as the product. Thus, protein oxidation in the ER is interlinked with $H_2O_2$ generation. Cells with high needs for de novo protein synthesis, such as cancer cells, generate abundant levels of ER-generated $H_2O_2$.

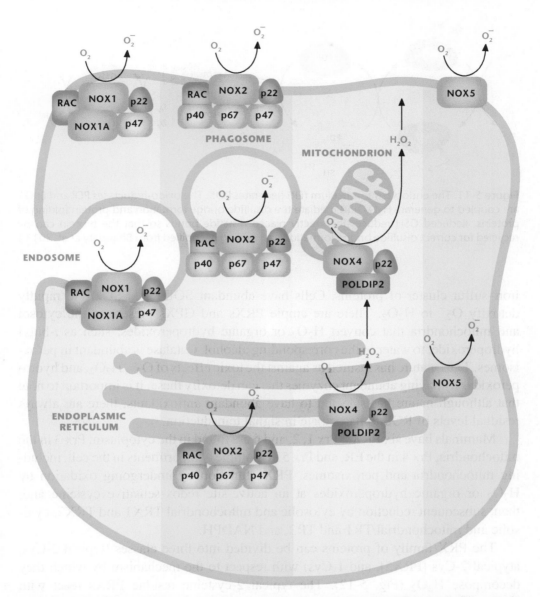

**Figure 5-10.** NADPH oxidases (NOXs) produce ROS. The NOX family of proteins localizes to the plasma, endoplasmic reticulum, and mitochondrial membranes. NOX catalytic subunits transfer electrons from NADPH across the membranes to molecular oxygen to generate $O_2^-$ superoxide or $H_2O_2$ hydrogen peroxide. (Adapted, with permission, from Drummond et al. 2011, © Macmillan.)

## NADPH IS USED TO DETOXIFY ROS

Given the reactivity and toxicity of ROS at high levels and that specific quantities of ROS determine various cellular signaling events, spatial and temporal regulatory strategies must exist to regulate intracellular ROS levels. $O_2^-$ can damage the

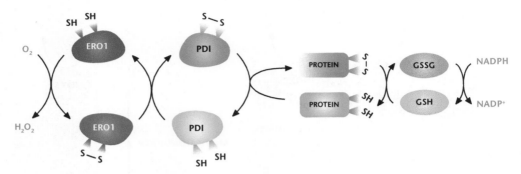

**Figure 5-11.** The endoplasmic reticulum (ER) generates ROS. The oxido-reductases PDI and ERO1 are coupled to generate $H_2O_2$ and mediate the disulfide bridge formation and protein folding of proteins. Reduced GSH reduces incorrectly placed disulfide bonds so that the protein can be recycled for correct disulfide bridge formation and folding. (Modified from Bhandary et al. 2013.)

iron–sulfur cluster of proteins. Cells have abundant SOD1 and SOD2 to rapidly detoxify $O_2^-$ to $H_2O_2$. There are ample PRXs and GPXs present in the cytosol and mitochondria that convert $H_2O_2$ or organic hydroperoxides, such as *t*-butyl hydroperoxide, to water or the corresponding alcohol. Catalase is abundant in peroxisomes. Thus, nature has insured us against the toxic effects of $O_2^-$, $H_2O_2$, and hydroperoxides by having abundant enzymes that can detoxify these. It is important to note that although nature has selected to have abundant antioxidants, there are always residual levels of ROS to participate in signal transduction.

Mammals have six PRXs: Prx 1, 2, and 6 are found in the cytoplasm, Prx 3 in the mitochondria, Prx 4 in the ER, and Prx 5 in multiple compartments in the cell, including mitochondria and peroxisomes. PRXs function by undergoing oxidation by $H_2O_2$ or organic hydroperoxides at an active-site redox-sensitive cysteine and, then, subsequent reduction by cytosolic and mitochondrial TRX1 and TRX2, cytosolic and mitochondrial TR1 and TR2, and NADPH.

The PRX family of proteins can be divided into three classes (typical 2-Cys, atypical 2-Cys [PRX5], and 1-Cys) with respect to the mechanism by which they decompose $H_2O_2$ (Fig. 5-12). The typical 2-cysteine residue PRXs react with $H_2O_2$ to form a $SO^-$ intermediate, which condenses with the resolving cysteine of the adjacent PRX to form an intermolecular dimer. The oxidized PRX dimer is reduced by TRX to complete the catalytic cycle. High levels of $H_2O_2$ can also react with the $SO^-$ intermediate to form a hyperoxidized $SO_2^-$. The $SO_2^-$ form of Prx is slowly reduced back to the native form of the enzyme by sulfiredoxin (SRX). The atypical 2-Cys Prxs have the same mechanism as typical 2-Cys PRXs, but they are functionally monomeric. The redox-sensitive cysteine and its corresponding resolving cysteine are located on the same polypeptide to generate an intermolecular disulfide bond. To recycle the disulfide, known atypical 2-Cys PRXs appear to use Trx as an electron donor. The 1-Cys Prxs contain the redox-sensitive cysteine without a

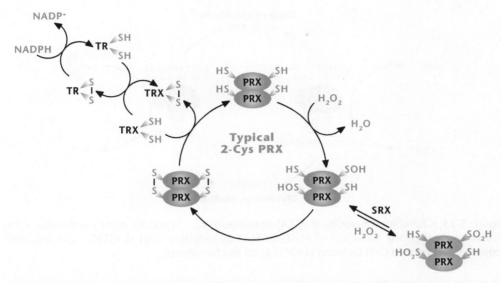

**Figure 5-12.** Peroxiredoxins (PRXs) scavenge $H_2O_2$. The peroxiredoxin (PRX1–6) family of proteins can be divided into three classes (typical 2-Cys, atypical 2-Cys [PRX5], and 1-Cys). PRXs function by undergoing oxidation by $H_2O_2$ at an active-site redox-sensitive cysteine and then subsequent reduction by cytosolic and mitochondrial thioredoxins (TRX1 and TRX2), cytosolic and mitochondrial thioredoxin reductase (TR1 and TR2), and NADPH. (Modified, with permission, from D'Autréaux and Toledano 2007, © Macmillan.)

corresponding resolving cysteine. Their mechanism of reacting with peroxide and their subsequent regeneration is not fully understood.

There are eight mammalian GPXs known to catalyze the reduction of $H_2O_2$ or organic hydroperoxides to water or the corresponding alcohols, respectively (Fig. 5-13). Reduced GSH (γ-L-glutamyl-L-cysteinyl-glycine) is used as the reducing agent. Mammalian GPX1–4 are selonproteins with a selenocysteine as the catalytic moiety that allows for a rapid reaction with peroxide or hydroperoxide and fast reducibility by GSH. This generates GSSG, often called oxidized GSH, which is reduced by GSH reductase. GPXs 5–8 use a cysteine residue that participates in detoxifying peroxides. GPX1 is ubiquitously expressed and found in both the cytosol and mitochondria. GPX2 is found in the intestinal epithelium, whereas GPX3 is secreted into plasma. GPX3 is highly expressed in adipose tissues. GPX4 is alternatively spliced and exists in three different isoforms: a cytosolic, a mitochondrial, and sperm nuclear GPX4. GPX5 and GPX6 are abundant in the epididymis and olfactory epithelium, respectively. GPX7 and GPX8 are cysteine-based with low-catalytic activity and presumed to function in the ER.

The function of these antioxidant enzymes is dependent on how fast they react with $H_2O_2$ (rate constant) and the expression of enzyme. The expression of these antioxidants is transcriptionally regulated by NRF2 (NF-E2 p45-related factor). PRXs are in high abundance and, therefore, are thought to be responsible for scavenging

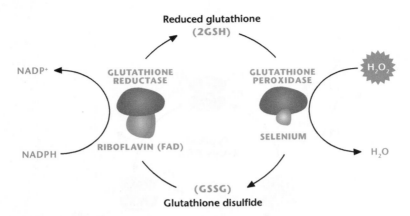

**Figure 5-13.** Glutathione peroxidase (GPX) detoxifies $H_2O_2$. There are eight mammalian GPXs known to catalyze the reduction of $H_2O_2$ to $H_2O$ by oxidizing GSH to GSSG. GSH reductase reduces GSSH back to GSH by using NADPH as an electron donor.

nanomolar levels of $H_2O_2$ associated with signaling. GPXs have higher rate constants than PRXs, but are less abundant and, therefore, are likely only important at higher intracellular concentrations of $H_2O_2$ when GPXs can begin to compete with PRXs for substrate. Therefore, it is possible that PRXs are critical for turning ROS signaling off, whereas GPXs are critical for buffering high levels of ROS to bring them to a level at which the cell evades damage and can initiate signaling stress responses.

## ARE ELEVATED ROS A CAUSE OR CONSEQUENCE OF PATHOLOGY?

A scientific colleague and friend once quipped, "if you don't have a mechanism just say it is ROS." Unfortunately, there is some truth to this statement. Many studies use *N*-acetylcysteine (NAC) as a putative antioxidant to show ROS as the causal agent in their biological system. In fact, over the past couple of decades, aberrant intracellular ROS levels and the cell's inability to clear the oxidants have been implicated in various diseases, including cancer, neurodegenerative disease, cardiovascular disease, diabetes, and gastrointestinal disease. But, a critical question here is whether ROS are causal or a consequence of these pathologies. Although the answer to this question is far from being answered, antioxidants are the most widely used or abused drugs worldwide. Perhaps, the most egregious is the overuse of vitamin C to prevent the common cold. The eminent chemist Linus Pauling was infatuated with the notion that megadoses of vitamin C would cure the ills of the world. He took abundant amounts of vitamin C and lived to the ripe old age of 93; he also captured the imagination of the public with his advocacy of vitamin C. However, repeatedly clinical trials have shown no efficacy of vitamin C in fighting the common cold. Yet, the myth continues, even among my own family members. Multiple clinical trials have uniformly failed to show beneficial effects of antioxidants on a variety of pathologies, including cancer. This suggests that either ROS are not causal for pathologies or we have not developed proper antioxidants. Another plausible explanation is that ROS at low levels regulate signaling to control physiological processes and adaptation to stressful conditions and, thus, antioxidants might interfere with our body's way to deal with stressful conditions. There are data to indicate in certain clinical trials that antioxidants have adverse effects, such as the NAC trial in patients suffering from the inflammatory syndrome sepsis caused by bacterial or viral infection. Further understanding of the importance of ROS in normal physiological processes and rationally designed antioxidants that do not undermine normal physiology, but might be effective under pathological conditions, are both needed.

---

**BOX 5-2. THE FREE RADICAL THEORY OF AGING**

Aging is an inevitable, universal phenomenon observed in all living organisms, yet its cause is open to debate. Some argue that aging is an effect of accumulation of damage to proteins, DNA, and lipids, and others suggest certain genes preprogram aging. In the 1950s, Denham Harman proposed the "mitochondrial free radical theory of aging" as a molecular explanation for why aging occurs. This theory is that free radicals, as by-products of mitochondrial oxidative metabolism, cause cumulative cellular damage that results in overall loss of organismal fitness over time. In subsequent decades, excessive ROS production has been postulated to be a causal agent for a variety of diseases, including diabetes, cancer, inflammatory disorders, and neurodegeneration. If this theory is correct, then increasing antioxidant capacity should increase organismal life span and ameliorate these devastating diseases. Yet, a large number of clinical trials have failed to show beneficial effects of antioxidants for these pathologies, and a handful of trials have suggested that antioxidants increase mortality. Worse, antioxidants have been shown to increase the progression of certain cancers. This indicates that either we have not used the right antioxidants or the mitochondrial free radical theory has some gaps. One possibility is that mitochondrial ROS are required to maintain homeostasis and adaptation to stress. Thus, increasing antioxidant capacity would limit adaptation to stress incurred during aging. Interestingly, recent findings show that low levels of mitochondrial ROS activate stress responses that are beneficial to the organism and extend life span. Going forward, if these findings are consistently upheld, then it would make the public rethink the overuse of antioxidants as antiaging agents.

---

## REFERENCES

Berg JM, Tymoczko JL, Stryer L. 2012. *Biochemistry*, 7th ed. WH Freeman, New York.

Bhandary B, Marahatta A, Kim HR, Chae HJ. 2013. An involvement of oxidative stress in endoplasmic reticulum stress and its associated diseases. *Int J Mol Sci* **14:** 434–456.

D'Autréaux B, Toledano MB. 2007. ROS as signalling molecules: Mechanisms that generate specificity in ROS homeostasis. *Nat Rev Mol Cell Biol* **8:** 813–824.

Drummond GR, Selemidis S, Griendling KK, Sobey CG. 2011. Combating oxidative stress in vascular disease: NADPH oxidases as therapeutic targets. *Nat Rev Drug Discov* **10:** 453–471.

Finkel T. 2011. Signal transduction by reactive oxygen species. *J Cell Biol* **194:** 7–15.

## ADDITIONAL READING

Brand MD. 2010. The sites and topology of mitochondrial superoxide production. *Exp Gerontol* **45:** 466–472.

Hekimi S, Lapointe J, Wen Y. 2011. Taking a 'good' look at free radicals in the aging process. *Trends Cell Biol* **21:** 569–576.

Janssen-Heininger YM, Mossman BT, Heintz NH, Forman HJ, Kalyanaraman B, Finkel T, Stamler JS, Rhee SG, van der Vliet A. 2008. Redox-based regulation of signal transduction: Principles, pitfalls, and promises. *Free Radic Biol Med* **45:** 1–17.

Lewis CA, Parker SJ, Fiske BP, McCloskey D, Gui DY, Green CR, Vokes NI, Feist AM, Vander Heiden MG, Metallo CM. 2014. Tracing compartmentalized nadph metabolism in the cytosol and mitochondria of mammalian cells. *Mol Cell* doi: 10.1016/j.molcel.2014.05.008.

Lyssiotis CA, Son J, Cantley LC, Kimmelman AC. 2013. Pancreatic cancers rely on a novel glutamine metabolism pathway to maintain redox balance. *Cell Cycle* **12:** 1987–1988.

Riemer J, Bulleid N, Herrmann JM. 2009. Disulfide formation in the ER and mitochondria: Two solutions to a common process. *Science* **324:** 1284–1287.

Ristow M. 2014. Unraveling the truth about antioxidants: Mitohormesis explains ROS-induced health benefits. *Nat Med* **20:** 709–711.

Winterbourn CC, Hampton MB. 2008. Thiol chemistry and specificity in redox signaling. *Free Radic Biol Med* **45:** 549–561.

# 6

# Carbohydrates

CARBOHYDRATES ARE THE MOST abundant macromolecules on our planet, in part because of the plant carbohydrates cellulose and starch, both composed of multiple conjugated glucose molecules. Cellulose is an important structural element of plant cell walls. Animals lack enzymes that can break down the cellulose into smaller glucose molecules, but they can break down starch into smaller glucose molecules. Animals also have glycogen, another carbohydrate composed of multiple conjugated glucose molecules. Many of us who exercise or play sports know that carbohydrates serve as a really good source of fuel during these strenuous endeavors. Unfortunately, most of us realize that overconsumption of carbohydrates can easily help us put on weight under nonexercise conditions. So, we know that carbohydrates can either be catabolized for energy (ATP) or used for anabolic functions, such as production of fatty acids.

Carbohydrates are divided into three major groups based on their structures: (1) simple sugars (monosaccharides and disaccharides), such as glucose or sucrose (glucose and fructose); (2) complex carbohydrates, such as glycogen, starch, and cellulose, which are multiple conjugated glucose molecules; and (3) glycoconjugates, which are modified forms of glucose covalently attached to either proteins (glycoproteins) or lipids (glycolipids), which participate in important functions, such as immunity, and as components of cell membranes. This chapter covers all three groups and highlights their importance in maintaining physiological functions.

## QUICK GUIDE TO CARBOHYDRATES

- Simple sugars, such as glucose, fructose, and galactose, can enter glycolysis (see Chapter 3).

- Gluconeogenesis begins with mitochondrial oxaloacetate being converted to phosphoenolpyruvate (PEP) by either mitochondrial or cytosolic phosphoenolpyruvate carboxykinase (PEPCK) (Fig. 6-1).

- Glycerol, alanine, lactate, and glutamine are the major substrates for gluconeogenesis.

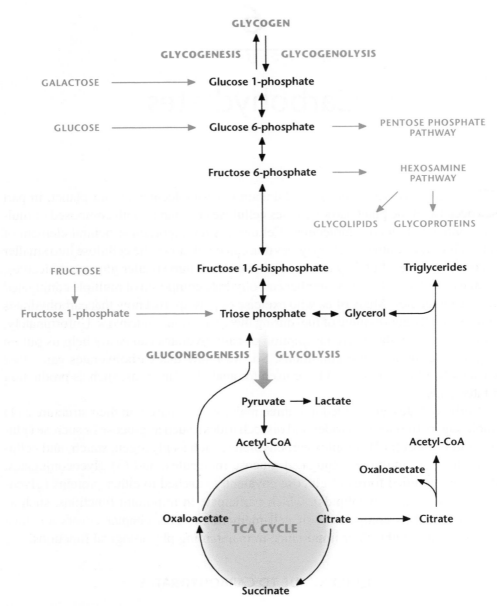

**Figure 6-1.** Overview of carbohydrate metabolism. Simple sugars, such as glucose, fructose, or galactose, have different points of entry into glycolysis. A process referred to as gluconeogenesis can also generate glucose. Complex carbohydrates such as glycogen can also enter glycolysis. The hexosamine pathway generates glycoproteins and glycolipids, which are modified forms of glucose covalently attached to either proteins (glycoproteins) or lipids (glycolipids), that participate in important functions in signaling and as components of cell membranes.

- There are three irreversible steps in glycolysis (hexokinase, phosphofructo-kinase-1 [PFK1], and pyruvate kinase) that are bypassed by enzymes specific to gluconeogenesis (glucose 6-phosphatase, fructose 1, 6-bisphosphatase, and phosphoenolpyruvate carboxykinase). All the enzymes that catalyze the reversible steps in glycolysis are used by gluconeogenesis.

- Glycogen can be degraded to glucose 1-phosphate to enter glycolysis (i.e., glycogenolysis). Conversely, glucose molecules can be converted into glucose 1-phosphate to generate glycogen (i.e., glycogenesis).

- Modified versions of glucose generated by the hexosamine pathway can modify proteins (i.e., glycoproteins) to change the activity or stability of proteins, thus linking glucose metabolism to cellular signaling.

## METABOLISM OF SIMPLE SUGARS

The Greek word "sakcharon" means sugar, and we use the word saccharide to denote a sugar. Simple sugars are monosaccharides, such as glucose, galactose or fructose; the disaccharides include lactose (galactose and glucose, milk sugar, sucrose (glucose and fructose, table sugar), and maltose (glucose and glucose) (Fig. 6-2). Sucrase

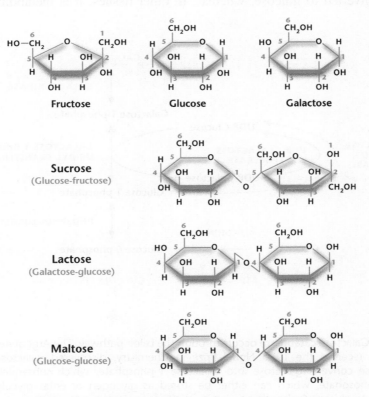

**Figure 6-2.** Monosaccharide and disaccharide structures.

and lactase are enzymes that break down sucrose and lactose into their monosacchar-ides, respectively (Fig. 6-2). Many adults are unable to metabolize lactose (i.e., they are lactose intolerant) usually because of diminished levels of the enzyme lactase. Certain bacteria in the colon use lactose as a source of fuel and, in the process, gen-erate methane ($CH_4$) and hydrogen gas ($H_2$), which cause discomfort in the gut and the embarrassing problem of flatulence.

Simple sugars have different levels of sweetness in mammals. The sensation of sweetness is based on sugars binding to G-protein-coupled receptors expressed on the surface of taste cells (gustatory cells) on our tongues, which stimulate a neuronal signal to brain. The differential affinity of sugars to the G-protein-coupled receptors in these cells determines the perceived sweetness. For example, fructose is sweeter than glucose, making certain fructose-based drinks addictive. Moreover, the meta-bolic fate of these sugars can be quite diverse. Glucose, galactose, and fructose enter glycolysis through different routes (Figs. 6-3 and 6-4). Glucose, as we learned in Chapter 3, becomes glucose 6-phosphate by an ATP-dependent reaction, using hex-okinases. Galactose enters through the Leloir pathway, in which galactokinase uses ATP to generate galactose 1-phosphate, which is converted to glucose 1-phosphate and, subsequently, to glucose 6-phosphate by the enzymes galactose-1-P-uridyl transferase and phosphoglucomutase, respectively. In the liver, glucose 6-phosphate can be converted to glucose, whereas, in other tissues, it is metabolized through

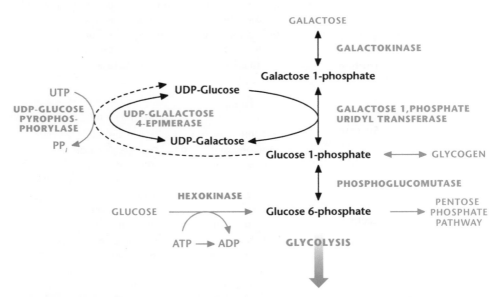

**Figure 6-3.** Galactose catabolism occurs through the Leloir pathway. The Argentine Luis Federico Leloir, who received the 1970 Nobel Prize in Chemistry, discovered galactose catabolism. Galactokinase converts galactose into galactose 1-phosphate, which subsequently becomes glucose 1-phosphate, which can either be stored as glycogen or enter glycolysis by being converted into glucose 6-phosphate.

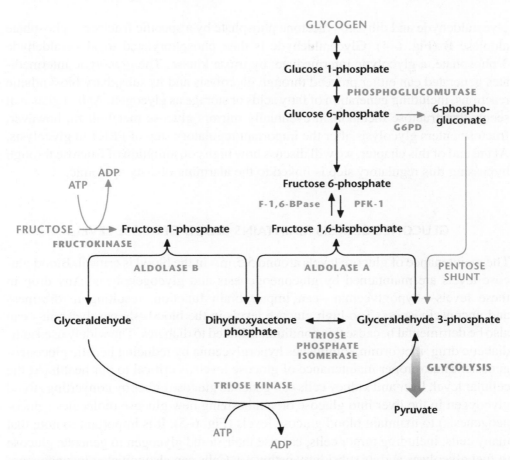

**Figure 6-4.** Fructose metabolism. Fructokinase converts fructose into fructose 1-phosphate, which subsequently is converted into glyceraldehyde and dihydroxyacetone phosphate by aldolase B that enters glycolysis. A key feature of fructose metabolism is that it bypasses the major regulatory step in glycolysis, the PFK1-catalyzed reaction.

glycolysis. The conversion of galactose to glucose 6-phosphate is slower than the rate by which glucose becomes glucose 6-phosphate. In proliferating cells, the replacement of glucose with galactose in vitro results in the galactose preferentially entering the pentose phosphate pathway (PPP) because mitochondrial oxidative phosphorylation provides ATP and the need for ribose 5-phosphate provided by the PPP is important for proliferation. In cells with mitochondrial oxidative phosphorylation defects, galactose metabolism through glycolysis is too slow to generate enough ATP to meet metabolic demands, resulting in metabolic catastrophe and cell death. Mitochondrial biologists use galactose sensitivity to determine whether a genetic mutation or pharmacologic inhibitor is suppressing oxidative phosphorylation.

Fructose is primarily metabolized by the liver and, to a lesser extent, by the small intestine and kidney. The first step is the phosphorylation of fructose to fructose 1-phosphate by fructokinase. Subsequently, fructose 1-phosphate is cleaved into glyceraldehyde and dihydroxyacetone phosphate by a specific fructose 1-phosphate aldolase B (Fig. 6-4). Glyceraldehyde is then phosphorylated to glyceraldehyde 3-phosphate, a glycolytic intermediate, by triose kinase. The glycolytic intermediates generated can either proceed through glycolysis and its subsidiary biosynthetic reactions, including generation of fatty acids or storage as glycogen. At first glance, it seems that fructose metabolism eventually mirrors glucose metabolism; however, fructose enters glycolysis after the important regulatory step of PFK1 in glycolysis. At the end of this chapter, we will discuss how high consumption of fructose through bypassing this regulatory step is linked to the alarming obesity epidemic.

## GLUCONEOGENESIS MAINTAINS BLOOD GLUCOSE LEVELS

The maintenance of glucose levels around 5.5 mM in the blood is critical. Blood glucose levels are maintained by gluconeogenesis and glycogenolysis. Any drop in these levels—hypoglycemia—can impair brain function, resulting in dizziness and unconsciousness. Too-high glucose levels in the blood—hyperglycemia—can also be detrimental because this condition is linked to diabetes. The widely used anti-diabetic drug, metformin, diminishes hyperglycemia by reducing hepatic gluconeogenesis. Thus, proper maintenance of glucose levels is critical to our health. At the cellular level, liver and kidney cells can generate glucose either by converting stored glycoygen in the liver into glucose or synthesizing new glucose molecules (gluconeogenesis) to maintain blood glucose levels (Fig. 6-5). It is important to note that many cells, including tumor cells, can use their stored glycogen to generate glucose to fuel glycolysis and its subsidiary pathways. Cells can also initiate gluconeogenesis to generate glycolytic intermediates that can go into subsidiary pathways, if needed, to generate macromolecules, such as lipids.

Gluconeogenesis primarily occurs in the liver and, to a lesser degree, in the kidney, in which the newly synthesized glucose is exported into circulating blood to provide glucose to vital organs, such as the brain, as well as red blood cells that derive their ATP solely from glucose-dependent glycolysis. Gluconeogenesis reactions occur both in the mitochondrial matrix and cytosol. In mammals, important sources that provide the carbons for gluconeogenesis are lactate, glycerol, and the amino acids alanine and glutamine. Lactate is generated by muscle and transported to the liver, in which it is converted into pyruvate to enter the gluconeogenesis. This is referred to as the Cori cycle (see Box 6-1).

As discussed in Chapter 3, there are three irreversible steps in glycolysis. These steps have to be bypassed for gluconeogenesis to proceed. The first step is the generation of PEP from pyruvate (Fig. 6-6). Pyruvate in the mitochondrial matrix is converted into oxaloacetate by the enzyme pyruvate carboxylase. This enzyme requires biotin as a cofactor and bicarbonate ($HCO_3$) as a substrate. The reaction is thermodynamically unfavorable and coupled to the Gibbs free energy provided by

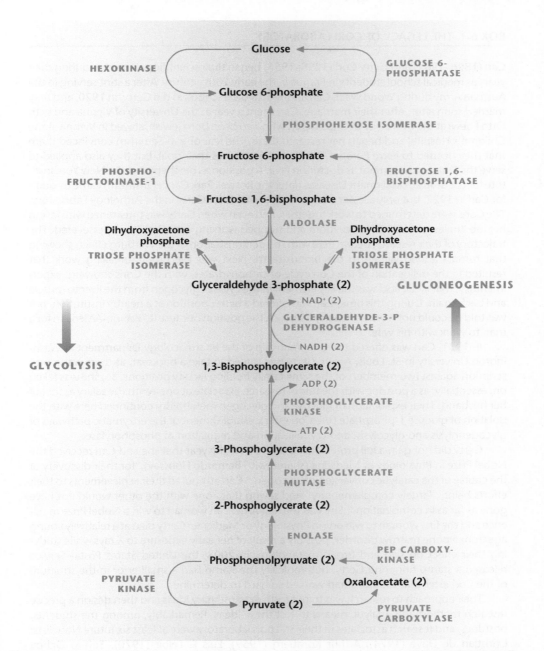

**Figure 6-5.** Gluconeogenesis. Glycolysis and gluconeogenesis share many enzymes; however, there are three irreversible reactions in glycolysis that have to be bypassed so that gluconeogenesis can ensue. The first reaction is the generation of PEP from pyruvate requiring pyruvate carboxylase and PEP carboxykinase. The second reaction is the conversion of fructose 1,6-bisphosphate to fructose 6-phosphate by F-1,6-BPase. The third reaction is the conversion of glucose 6-phosphate to glucose by glucose 6-phosphatase.

## BOX 6-1. THE LEGACY OF CORI LABORATORY

Carl (1896–1984) and Gerty Cori (1896–1957) began their scientific partnership during their years as medical school students in Prague in the early 20th century. After a stint serving in the Austrian Army during World War I, Carl finished medical school, as did Gerty, in 1920, and they married soon after. After their marriage, Carl spent a year at the University of Vienna and with Otto Loewi at the University of Graz. Gerty, who had been born Jewish, stayed in Vienna at the Children's Hospital and began her research there. The fear of anti-Semitism convinced them that they needed to leave Europe. The United States was their goal, but they also applied to serve the Dutch government as doctors in Java. A position as biochemist at the New York Institute for the Study of Malignant Diseases (later the Roswell Park Cancer Institute) came through for Carl in 1922, but only a lesser position was available for Gerty in the Pathology Laboratory. The Coris were determined to work together, and even when Gerty was threatened with losing her job "unless she stayed in her room and stopped working with Carl" they persevered. The trajectory of their research began here with the demonstration of the Warburg effect, showing that tumors added lactate to the bloodstream. Next was the groundbreaking work that resulted in the delineation of the Cori cycle of carbohydrates, with the Coris showing, experimentally, that lactic acid was the key element in the cycle of glycogen from the liver to muscle and back again. During this time, Carl was offered a better position at a nearby institution, but was told he could not work with Gerty if he took the position because it "was un-American for a man to work with his wife."

In 1931, Carl was offered the Chairmanship of the Pharmacology Department at Washington University in St. Louis. Again, Gerty was forced to take a backseat, as there was a proscription against two members of the same family holding faculty positions. So, she was taken on, essentially, as a postdoc with the title of research associate at one-tenth the salary as that of her husband. Their exploration of glucose and glycogen metabolism continued here with the isolation of glucose 1-phosphate (the Cori ester), establishment of the enzymatic pathways of glycogenolysis and glycolysis, and crystallization and regulation of phosphorylase.

Gerty did not gain a full professorship until 1947, the year that she and Carl received the Nobel Prize in Physiology or Medicine (shared with Bernardo Houssay) "for their discovery of the course of the catalytic conversion of glycogen." Carl attributed their achievements to their efforts being "largely complementary" and saying that "one with the other would not have gone as far as in combination." She was the first American woman to win a Nobel Prize in science and the first woman to win one in Physiology or Medicine. Gerty died at a relatively young age from a bone marrow disorder, possibly a result of her early exposure to X rays while studying their effect on skin and organ metabolism. In 2008, the United States Postal Service released a stamp honoring Gerty, but ironically the stamp had a small error in the structure of the Cori ester that the Coris had worked so hard to determine.

Their approach to research was to put forth extraordinary ideas and then design a precise research method and analytic means to test these ideas. Remarkably, among the students, postdocs, and research associates in their St. Louis laboratory were at least six future Nobelists: Christian de Duve (1974), Arthur Kornberg (1959), Luis F. Leloir (1970), Severo Ochoa (1959), Earl W. Sutherland (1971), and Edwin G. Krebs (1992). The Cori's identification of the "PR enzyme" in the conversion of phosphorylase *a* to phosphorylase *b* was later found to be the first protein phosphatase to be identified (PP1 of the PPP phosphatase family). This discovery led to Edmond Fischer and Edwin G. Krebs showing that the phosphorylase *b* to phosphorylase *a* conversion involved phosphorylation, which turned out to be a broader method for regulating protein function.

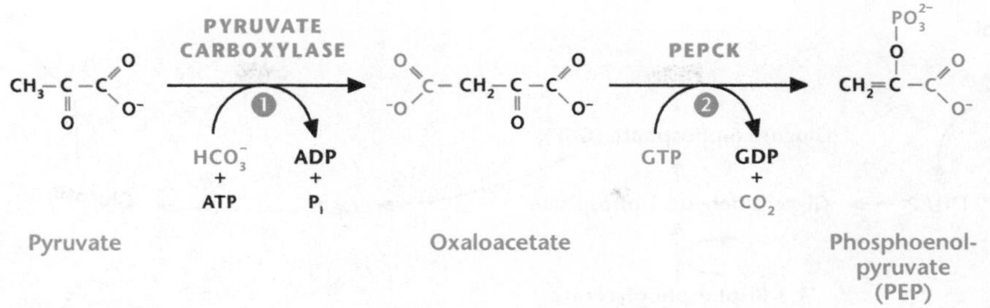

**Figure 6-6.** Pyruvate conversion into PEP.

converting ATP to ADP. Acetyl-CoA is a positive allosteric regulator of pyruvate carboxylase. Therefore, if acetyl-CoA levels increase, then acetyl-CoA stimulates pyruvate carboxylase to generate oxaloacetate, and these two metabolites could make citrate to initiate TCA cycle. However, if the liver cells' energy charge is not low, they can convert the oxaloacetate into PEP by PEPCK by coupling this reaction to the conversion of GTP to GDP (Fig. 6-6).

Human liver cells have two distinct PEPCK genes that encode cytosolic and mitochondrial matrix enzymes. Gluconeogenic amino acid alanine is converted into pyruvate and uses the cytosolic PEPCK, which converts cytosolic oxaloacetate to generate PEP (Fig. 6-7). In this pathway, the pyruvate in the mitochondria is converted into mitochondrial oxaloacetate by pyruvate carboxylase. Mitochondria do not have a mechanism to transport oxaloacetate. Thus, oxaloacetate must be converted into malate, which can be transported into the cytosol. This reaction is catalyzed by mitochondrial malate dehydrogenase 2. Subsequently, cytosolic malate dehydrogenase 1 (MDH1) oxidizes malate into cytosolic oxaloacetate by coupling to $NAD^+$ reduction to NADH. Once PEP is generated, it uses most of the glycolytic enzymes to eventually become glucose. The NADH generated by malate dehydrogenase 1 is used by glyceraldehyde 3-phosphate dehydrogenase (GAPDH) to convert 1,3-bisphosphoglycerate into glyceraldehyde 3-phosphate.

Lactate generated by muscle is also used as a gluconeogenic substrate through conversion into pyruvate. The enzyme lactate dehydrogenase converts lactate into pyruvate by coupling to $NAD^+$ conversion to NADH. Pyruvate becomes oxaloacetate by pyruvate carboxylase (PC). Oxaloacetate is converted into PEP in the mitochondrial matrix by PEPCK2 and, subsequently, is transported into the cytosol to enter gluconeogenesis. Note that oxaloacetate is not converted into malate in the mitochondria by malate dehydrogenase 2 (MDH2) and thus is not transported into the cytosol, in which it would be converted back into oxaloacetate through coupling the reaction with the conversion of $NAD^+$ to NADH. The generation of lactate from pyruvate already generates NADH in the cytosol needed for GAPDH reaction, thus alleviating the necessity of malate shuttling out of the mitochondria to generate NADH.

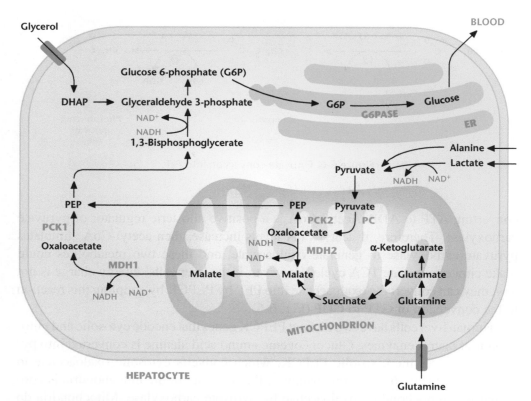

**Figure 6-7.** Multiple substrates feed into gluconeogenesis. Alanine, lactate, glycerol, and glutamine can generate glucose. Glycerol enters gluconeogenesis through conversion into dihydroxyacetone phosphate (DHAP), a reaction catalyzed by glycerol 3-phosphate dehydrogenase. Alanine, lactate, and glutamine have to be converted into oxaloacetate, which enters gluconeogenesis through conversion into PEP by phosphoenolpyruvate carboxykinase.

Once PEP goes through reverse glycolysis, there are two steps of glycolysis that are not reversible: those catalyzed by PFK1 and hexokinase. The corresponding enzymes that catalyze the reverse reactions are fructose 1,6-bisphosphatase(F-1, 6-BPase) and glucose 6-phosphatase, respectively (Fig. 6-5). Glycerol can also contribute to gluconeogenesis by the conversion of glycerol to glycerol 3-phosphate by glycerol kinase. Subsequently, glycerol 3-phosphate becomes the glycolytic intermediate dihydroxyacetone phosphate by mitochondrial glycerol 3-phosphate dehydrogenase. Dihydroxyacetone phosphate is converted into glyceraldehyde 3-phosphate, which eventually becomes glucose.

Gluconeogenesis is an endergonic process (requires energy) when glycerol, alanine, and lactate are substrates. Glycerol, alanine, and lactate entry does not generate ATP. Moreover, the conversion of pyruvate to oxaloacetate uses ATP and gluconeogenesis, through reversal of glycolytic steps, also consumes ATP (Fig. 6-7). However, glutamine gluconeogenesis is unique in that it represents an exergonic reaction. Glutamine through glutaminolysis (see Chapter 4) becomes α-ketoglutarate, which goes through the TCA cycle to ultimately produce malate, which shuttles into the cytosol to enter gluconeogenesis. Entry of glutamine into the TCA cycle generates GTP, NADH, and $FADH_2$ in the mitochondrial matrix that produces ATP to drive gluconeogenesis in the cytosol.

## REGULATION OF GLUCONEOGENESIS

It is important to realize that gluconeogenesis is a tightly regulated pathway that does not allow cells to simultaneously conduct glucose degradation by glycolysis and glucose synthesis by gluconeogenesis. There is reciprocal control of these pathways to prevent a futile cycle (Fig. 6-8). A key regulatory step is how PFK1 and F-1,6-BPase are reciprocally regulated by AMP, citrate, and fructose 2,6-bisphosphate (F-2,6-BP). If the energy charge decreases in cells, then AMP levels increase, leading to PFK1 activation (increasing glycolytic flux) and inhibition of F-1,6-BPase (decreasing gluconeogenic flux). In contrast, if citrate levels build up in the cytosol because the TCA cycle is backed up, then glycolytic flux is reduced through citrate inhibition of PFK1. Simultaneously, gluconeogenic flux is increased through citrate activation of F-1,6-BPase. The third metabolite and most potent allosteric regulator of glycolysis and gluconeogenesis is F-2,6-BP, which is generated by phosphofructokinase-2 (PFK2) and degraded by fructose 2,6-bisphosphatase (F-2,6-BPase). F-2,6-BP activates PFK1 and inhibits F-1,6-BPase. A single protein contains both PFK2 and F-2,6-BPase activities. The interconversion of PFK2 and F-2,6-BPase is achieved by cAMP-dependent protein kinase A (PKA) phosphorylation of PFK2 to produce F-2,6-BPase. Thus, stimuli that increase cAMP, such as the hormone glucagon, promote gluconeogenesis (see Box 6-2).

## GLYCOGEN SYNTHESIS AND DEGRADATION MAINTAINS GLUCOSE HOMEOSTASIS

Glycogen is a large, highly branched polysaccharide consisting of individual glucose molecules joined by $\alpha$-(1,4) and $\alpha$-(1,6) glycosidic bonds. Glycogen is degraded and synthesized in the cytosol, notably in liver and muscle cells, but also in other cells, including tumor cells and cells in the retina. The key enzymes are glycogen synthase, glycogen phosphorylase, and branching/debranching enzymes. Glycogen synthesis from glucose is performed by the enzyme glycogen synthase, which uses UDP-glucose and glycogen as substrates. The enzyme UDP-glucose pyrophosphorylase exchanges the phosphate on C-1 of glucose 1-phosphate for UDP to generate UDP-glucose. The energy of the phospho–glycosyl bond of UDP-glucose is used by glycogen synthase to catalyze the incorporation of glucose into glycogen (Fig. 6-9). UDP is, subsequently, released from the enzyme. The $\alpha$-1,6 branches

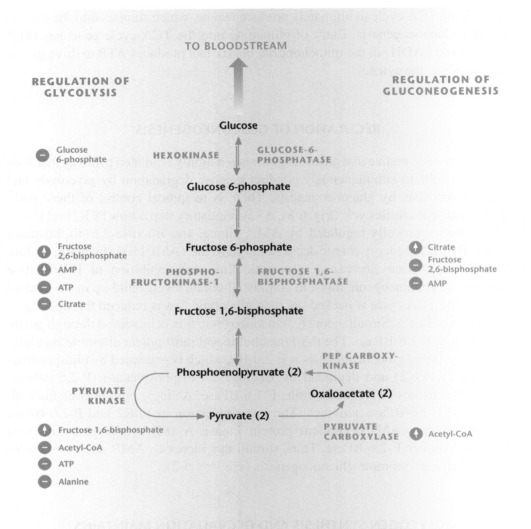

**Figure 6-8.** Reciprocal regulation of glycolysis and gluconeogenesis. PFK1 and fructose 1,6-bisphosphate (F-2,6-BPase) are key regulatory enzymes in glycolysis and glucogenogensis, respectively. AMP and F-2,6-BP activate PFK1 and inhibit F-2,6-BPase.

in glucose are produced by amylo-(1,4–1,6)-transglycosylase, also termed the branching enzyme.

Glycogen phosphorylase degrades stored glycogen by the process of glycogenolysis. Single glucose residues are removed phosphorolytically from α-(1,4)-linkages within the glycogen, generating the product glucose 1-phosphate. Glycogen phosphorylase cannot remove glucose residues from the branch points (α-(1,6) linkages) in glycogen, thus, requiring debranching enzymes. The phosphorylated form of glucose removed from glycogen does not require ATP hydrolysis, and the

---

**BOX 6-2. MAINTAINING GLUCOSE HOMEOSTASIS DURING FED–FAST CYCLE**

The fed–fast cycle starts nightly after our evening meals (fed state) followed by nightly sleep (fast state). Throughout this cycle, blood glucose levels have to be maintained. The cycle has fluctuations in metabolic hormones insulin and glucagon, which help maintain blood glucose levels. After a meal, the increase in glucose levels quickly triggers secretion of insulin by the pancreas, which suppresses liver gluconeogenesis. Insulin activates glycogen synthase and inactivates glycogen phosphorylase, resulting in liver glycogen synthesis. Insulin also stimulates glucose uptake in the muscle and adipose tissue for storage. Collectively, these actions of insulin lower blood glucose levels. Several hours after a meal, the blood glucose levels begin to decrease, leading to a decrease in insulin secretion and an increase in glucagon secretion from the pancreas. The decrease in insulin levels diminishes uptake of glucose by muscle and adipose tissue, contributing to the maintenance of the blood glucose level. Glucagon prevents glycogen synthesis and stimulates glycogen breakdown in the liver by activating glycogen phosphorylase and inactivating glycogen synthase. In addition, glucagon increases the production of cAMP to activate PKA, which converts PFK2 to F-2,6-BPase to suppress glycolysis and stimulate gluconeogenesis in the liver. Once we wake up and eat breakfast, glucagon levels rapidly diminish and insulin levels increase, causing cAMP to be degraded and PKA to be inactivated. This causes the conversion of F-2,6-BPase to PFK2 to activate PFK1 to stimulate glycolysis and inhibit F-1,6-BPase to repress gluconeogenesis. Thus, hormonal control of F-2,6-BP rapidly regulates glycolysis and gluconeogenesis.

---

equilibrium of the reaction is favorably driven by the high concentration of $P_i$ in cells. Subsequently, the glucose 1-phosphate is converted into glucose 6-phosphate, which can either precede glycolysis or the PPP, by phosphoglucomutase (Fig. 6-9). The release of a phosphorylated glucose molecule from glycogen also ensures that the glucose residue does not freely diffuse from cells. This is particularly important in muscle cells, in which the glucose residues generated from glycogenolysis is needed to proceed through glycolysis for ATP generation. Muscle cells lack glucose 6-phosphatase, thus glucose 6-phosphate cannot be generated into free glucose molecules. In contrast, liver contains glucose 6-phosphatase, thus, allowing glucose 6-phosphate molecules generated from glycogen into free glucose to maintain blood glucose levels.

## INTERSECTION OF CARBOHYDRATES AND SIGNALING

Carbohydrates can also play an important role in signaling by posttranslationally modifying proteins. A number of oligosaccharides (glycans) attach covalently to a protein to alter protein stability and activity; these constructs are referred to as glycoproteins. The linkage of proteins to carbohydrates in glycoproteins is through either a *N*-glycosidic bond or *O*-glycosidic bond. The *N*-glycosidic linkage is through the amide group of asparagine to *N*-acetylglucosamine (GlcNAc). The *O*-glycosidic linkage is to the hydroxyl of threonine, serine, or hydroxylysine. The most common linkage to threonine or serine is *N*-acetylgalactosamine (GalNAc) by *O*-GlcNAc transferase (OGT). This modification occurs in the endoplasmic reticulum and Golgi

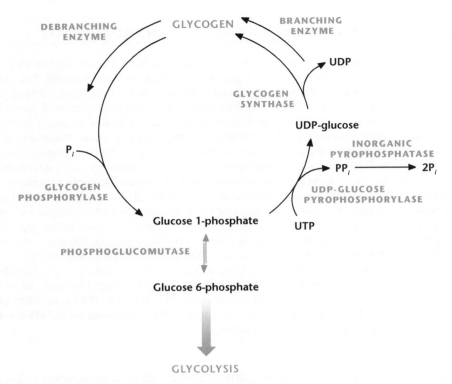

**Figure 6-9.** Glycogen metabolism. Glycogen phosphorylase breaks down glycogen into glucose 1-phosphate, whereas glycogen synthase synthesizes glucose 1-phosphate molecules into glycogen. Glucose 1-phosphate can be interconverted into glucose 6-phosphate by phosphoglucomutase.

apparatus. Many membrane-bound and secreted proteins are glycoproteins. These include many of the cell-surface receptors of the immune system, hormones, such as erythropoietin, and the mucins, which are secreted in the mucus of the respiratory and digestive systems. The predominant sugars found in glycoproteins are glucose, galactose, fucose, mannose, *N*-acetylneuraminic acid, GalNAc, and GlcNAc.

The details of all these different modifications are beyond the scope of this book. However, one important modification worth examining is the *O*-GlcNAc modification because it is regulated by glucose metabolism. Recall from Chapter 3 that glucose metabolism can proceed down to glycolysis to generate ATP or the carbons can be funneled into different subsidiary pathways, including the hexosamine pathway. The rate-limiting step in this pathway is the first step, catalyzed by glutamine fructose 6-phosphate amidotransferase (GFAT), which uses glutamine and fructose 6-phosphate to generate the product glucosamine 6-phosphate (Fig. 6-10). Subsequently, glucosamine 6-P-*N*-acetyltransferase (GNA) uses acetyl-CoA and glucosamine 6-phosphate as substrates to generate *N*-acetylglucosamine 6-phosphate, which is converted to *N*-acetylglucosamine 1-phosphate by phosphoacetylglucosamine mutase (PAGM). In the final step of the hexosamine

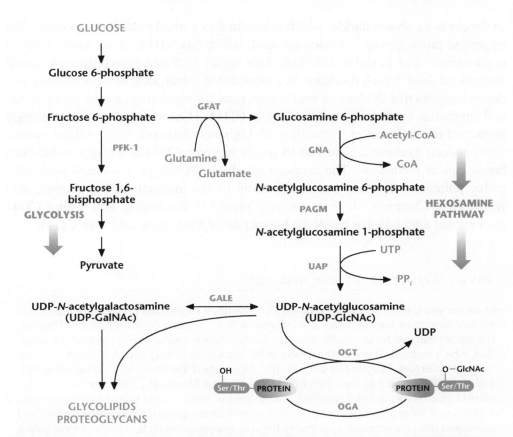

**Figure 6-10.** The hexosamine pathway generates glycoconjugates. Fructose 6-phosphate can be converted into glucosamine 6-phosphate by GFAT to initiate the hexosamine pathway, which, through a series of reactions, generates UDP-*N*-acetylglucosamine (UDP-GlcNAC) and *N*-acetylgalactosamine (UDP-GalNAC), which are used to generate glycolipids, proteoglycans, and glycoproteins. OGT uses UDP-GlcNAc to *O*-GlcNAcylate serine and threonine residues of proteins to modify their activity. These proteins can have their GlcNAc moiety removed by OGA.

pathway, UDP is added to *N*-acetylglucosamine-1-phosphate to generate UDP-GlcNAc by UDP-*N*-acetylglucosamine pyrophosphorylase (UAP). UDP-GlcNAc can be converted into UDP-GalNAc by UDP-galactose 4-epimerase. UDP-GlcNAc is used by OGT to *O*-GlcNAcylate proteins. These proteins can have their GlcNAc moiety removed by *O*-GlcNAcase (OGA). Several intracellular proteins, such as transcription factors and RNA polymerase II, can be modified by *O*-GlcNAc linkage. Furthermore, there is growing evidence linking the insulin-resistant phenotype observed in diabetes to increasing *O*-GlcNAcylation of proteins. An increase in UDP-GlcNAc levels inhibits GFAT activity through a negative feedback mechanism to reduce the flux through the pathway.

UDP-GlcNAc can also be used to generate proteoglycans, which are found in the extracellular matrix and connective tissue. The majority of the proteoglycan molecule is a polysaccharide, which is attached to a small protein component. One important proteoglycan is hyaluronic acid, which has 50,000 disaccharide units of D-glucuronic acid linked to GlcNAc. This highly hydrated proteoglycan is found in synovial fluid, which functions as a lubricant in joints. Recent accumulating evidence suggests that hyaluronic acid is important for regulating cellular proliferation and migration by binding to its receptor CD44. Hyaluronic acid is increasingly expressed in breast cancer stem cells with high metastatic potential, and the interaction between hyaluronic acid and its major receptor CD44 is thought to facilitate breast cancer metastasis. The invasive fibroblast phenotype associated with idiopathic pulmonary fibrosis is also dependent on the interaction of hyaluronic acid with CD44. Currently, there is research interest in developing and testing CD44 monoclonal antibodies to block the interaction of hyaluronic acid with CD44.

---

### BOX 6-3. IS FRUCTOSE THE NEW TOBACCO?

In recent years, fructose has become a much-maligned sugar. Robert H. Lustig, a pediatric endocrinologist at the University of California at San Francisco Benioff Children's Hospital, has made headlines for years with his public health crusade against excess fructose consumption, which he thinks is one of the drivers of the increase in obesity and diabetes. A YouTube video of his lecture, "Sugar: The Bitter Truth," has received more than four million views. Politicians, including former New York City Mayor Michael Bloomberg, have heeded his arguments to ban restaurant and concession stand sales of sugary drinks bigger than 16 ounces. A difficulty in linking fructose consumption to obesity over a period of time is that randomized controlled trials are impossible, primarily because everyone reverts to a more normal eating pattern after a couple of months.

But what is so bad about fructose or high-fructose corn syrup? Can it be turned into fat? Is table sugar different than high-fructose corn syrup? There are a few misconceptions regarding high-fructose corn syrup and table sugar and the ratio of fructose to glucose. Table sugar or sucrose has equal amounts of glucose and fructose (50%); high-fructose corn syrup is not that much different—it has 55% fructose. However, the big difference between glucose and fructose is how they are metabolized. For example, 100 calories of potato, which contains glucose molecules that exist in the form of starch, is metabolized very differently than 100 calories of table sugar, which contains 50% fructose and 50% glucose. The calories are the same, but all the cells in our body can use glucose in potatoes, whereas it is the liver that primarily uses fructose. The metabolism of glucose and fructose in the liver is quite different. Most cells do not have GLUT5, the major transporter of fructose. The liver has abundant GLUT5 transporters, making fructose readily accessible for metabolism. Interestingly, tumor cells can also metabolize fructose. Fructose enters the glycolysis pathway through conversion into fructose 1-phosphate by fructokinase. Subsequently, Aldolase B converts fructose 1-phosphate into glyceraldehyde and dihydroxyacetone phosphate.

As discussed in Chapter 3, a major regulatory step in glycolysis is phosphofructose kinase 1. This step is bypassed by fructose's entry into glycolysis. Thus, if the liver has met its energetic needs, it converts excess fructose into glyceraldehyde, which is converted into glycerol 3-phosphate, a precursor for triglycerides (Box 6-3, Fig. 1). Fructose can also generate dihydroxyacetone phosphate, which becomes glyceraldehyde 3-phoshpate and eventually goes through glycolysis and into the mitochondrial TCA cycle. The excess citrate generated is

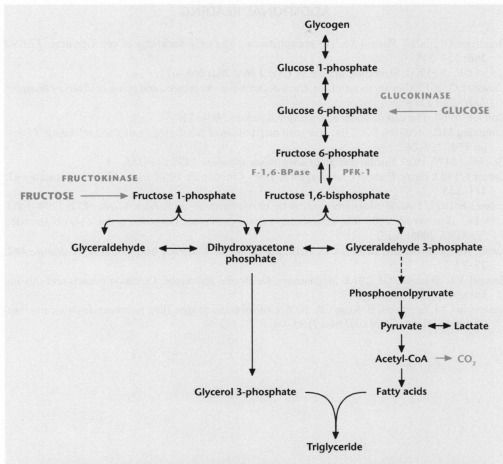

**Box 6-3, Figure 1.** Fructose generates triglycerides in the liver. Fructose enters the glycolysis pathway through conversion into fructose 1-phosphate by fructokinase. Subsequently, Aldolase B converts fructose 1-phosphate into glyceraldehyde and dihydroxyacetone phosphate. A major regulatory step in glycolysis is PFK1, which is bypassed by fructose's entry into glycolysis. Thus, if the liver has met its energetic needs, it converts excess fructose into glyceraldehyde, which is converted into glycerol 3-phosphate, a precursor for triglycerides. Fructose also generates dihydroxyacetone phosphate, which can become glyceraldehyde 3-phosphate and, eventually, go through glycolysis and into the mitochondrial TCA cycle. The excess citrate generated is exported to cytosol to generate fatty acids, another precursor to triglycerides.

exported to cytosol, where it is converted into acetyl-CoA to generate fatty acids, another precursor to triglycerides.

Thus, you can burn off the glucose in every cell, but the fructose is converted into fat in the liver, which causes insulin resistance. What about artificial sweeteners in diet sodas? A recent large study on more than 60,000 women found that diet drinks raised the risk of diabetes more than fruit juices or sugar-sweetened drinks over a 14-year period. Artificial sweeteners trick the body into thinking sugar is on its way and this causes the pancreas to pump out insulin, which can increase storage of nutrients, such as fat in the liver. Chapter 7 discusses more about fat metabolism.

## ADDITIONAL READING

Brautigan DL. 2013. Protein Ser/Thr phosphatases—The ugly ducklings of cell signaling. *FEBS J* **280:** 324–345.

Cahill GF, Jr. 1970. Starvation in man. *N Engl J Med* **282:** 668–675.

Cantley LC. 2013. Cancer, metabolism, fructose, artificial sweeteners, and going cold turkey on sugar. *BMC Biol* **12:** 8.

Cori CF. 1969. The call of science. *Annu Rev Biochem* **38:** 1–21.

Cumming MC, Morrison SD. 1960. The total metabolism of rats during fasting and refeeding. *J Physiol* **154:** 219–243.

Goldblatt MW. 1929. Insulin and gluconeogenesis. *Biochem J* **23:** 243–255.

Larner J. 1992. Gerty Theresa Cori: August 8, 1896–October 26, 1957. *Biogr Mem Natl Acad Sci* **61:** 111–135.

Leloir LF. 1971. Two decades of research on the biosynthesis of saccharides. *Science* **172:** 1299–1303.

Love DC, Hanover JA. 2005. The hexosamine signaling pathway: Deciphering the 'O-GlcNAc code.' *Sci STKE* **2005:** re13.

Lustig RH, Schmidt LA, Brindis CD. 2012. Public health: The toxic truth about sugar. *Nature* **482:** 27–29.

Samuel VT, Shulman GI. 2012. Mechanisms for insulin resistance: Common threads and missing links. *Cell* **148:** 852–871.

Taniguchi CM, Emanuelli B, Kahn CR. 2006. Critical nodes in signalling pathways: Insights into insulin action. *Nat Rev Mol Cell Biol* **7:** 85–96.

# 7

## Lipids

FAT GETS A BAD RAP BECAUSE IT IS IMPLICATED AS THE major culprit behind the obesity pandemic worldwide. The so-called fast-food diet that the United States has exported worldwide is blamed for the increase in body mass index (BMI), which is associated with health risks. Recent studies have emphasized that where you carry the fat matters in terms of health risks. For example, people carrying excess fat in the abdomen have higher risks of cancer and cardiovascular disease. It is clear that the biology of fat is complex. Fat, which is synonymous with lipids, plays an important role in our cells to maintain homeostasis. Lipids generate ATP and are involved in the synthesis of vitamins, hormones, bile salts, eicosanoids, and cellular membranes, as well as the regulation of cellular signaling. Cholesterol and phospholipids are essential components of membranes within the cell. The anabolism and catabolism of lipids is compartmentalized. Anabolism primarily occurs in the cytosol and endoplasmic reticulum, whereas the catabolism primarily occurs in mitochondria.

Lipids constitute an enormous topic and have many ramifications concerning human disease. This chapter will cover three basic aspects of lipid biology: (1) production of lipids, (2) catabolism of lipids to generate ATP, and (3) lipids as signaling molecules.

### QUICK GUIDE TO LIPIDS

- Lipids, such as triacylglycerol (TAG) and phospholipids, are generated from glucose-derived glycerol and mitochondrial-derived fatty acids (Fig. 7-1).

- Fatty acid synthesis takes place in the cytosol, where mitochondrial citrate serves as the precursor to eventually generate palmitate, which can be modified to other fatty acids.

- Fatty acid β-oxidation occurs in the mitochondrial matrix. Fatty acids are transported into the matrix through carnitine acyltransferase I (CPTI) located in the outer mitochondrial membrane, along with carnitine acyltransferase II (CPTII) and carnitine-acylcarnitine translocase, located in the inner mitochondrial membrane.

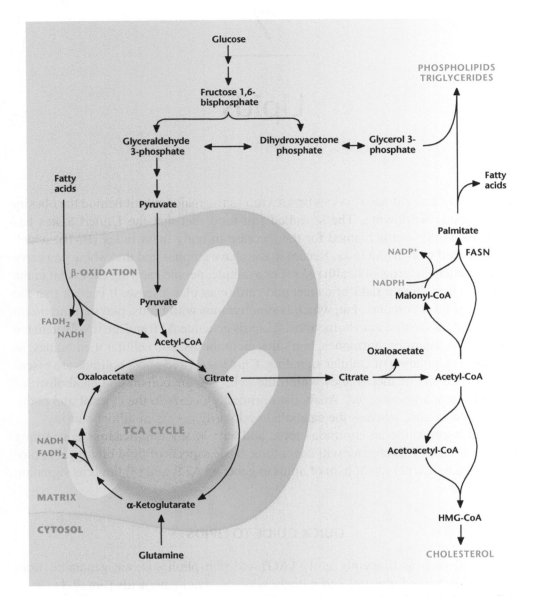

**Figure 7-1.** Overview of lipid metabolism. Lipid synthesis requires the glycolytic intermediate dihydroxyacetone phosphate and TCA cycle intermediate citrate to generate glycerol 3-phosphate and acetyl-CoA, respectively. Fatty acid synthase (FASN) converts acetyl-CoA to palmitate, which, together with glycerol 3-phosphate, generates lipids, such as triglycerides and phospholipids. Acetyl-CoA is also used to generate cholesterol. Lipids break down into fatty acids, which can be used by mitochondria in β-oxidation to generate ATP.

- Fatty acid synthesis is coupled to NADPH → NADP$^+$, whereas fatty acid oxidation generates acetyl-CoA, NADH, and FADH$_2$ to produce ATP through oxidative phosphorylation.

- Fatty acid synthesis is regulated by acetyl-CoA carboxylase (ACC), which is activated by citrate and inhibited by the fatty acid palmitate.

- Fatty acid β-oxidation is regulated by malonyl-CoA, which inhibits carnitine acyltransferase (CPTI) activity, thereby preventing fatty acid import into the mitochondrial matrix for β-oxidation.

- Lipids can modify proteins to alter their function. Notable modifications are *N*-myristoylation, *S*- or *N*-palmitoylation, and *S*-prenylation.

- Lipids, such as eicosanoids, phosphoinositides, and sphingolipids, serve as signaling molecules.

- The cholesterol biosynthetic pathway initiates in the cytosol and is controlled by the enzyme 3-hydroxy-3-methylglutaryl CoA reductase (HMG-CoA reductase), the target of statins (class of cholesterol-lowering drugs).

## GLYCOLYSIS AND MITOCHONDRIAL METABOLISM GENERATE LIPIDS

There are multiple types of lipids, including TAGs (commonly known as triglycerides) and phospholipids. TAG is a glycerol attached to mixture of saturated and unsaturated fatty acids (Fig. 7-2). Phospholipids are typically composed of two fatty acids linked to a glycerol and are attached to a polar molecule, like choline, via a phosphate group (Fig. 7-2). A close examination of the structure of TAGs and phospholipids informs us that there is a glucose-derived product (glycerol) attached to fatty acids, a long hydrocarbon chain composed of multiple acetyl groups, indicating that the two metabolic pathways involved in generating TAGs and phospholipids are glycolysis and TCA cycle. Fatty acid synthesis occurs in the cytosol, where there is a high NADPH/NADP$^+$ ratio to drive the reactions. The addition of glycerol to fatty acids to generate TAGs and phospholipids occurs in the endoplasmic reticulum.

Fatty acids are classified according to the number of carbon double bonds. Saturated fatty acids have no double bonds. Monounsaturated fatty acids have one double bond and polyunsaturated fatty acids have two or more double bonds. Some common saturated fatty acids are palmitic, butyric, and stearic acids; oleic and linoleic acids are mono- and polyunsaturated fatty acids, respectively (Fig. 7-2). Fatty acid synthesis combines eight two-carbon acetyl groups derived from mitochondrial citrate-generated acetyl-CoA to form a palmitate, a 16-carbon saturated fatty acid. Palmitate

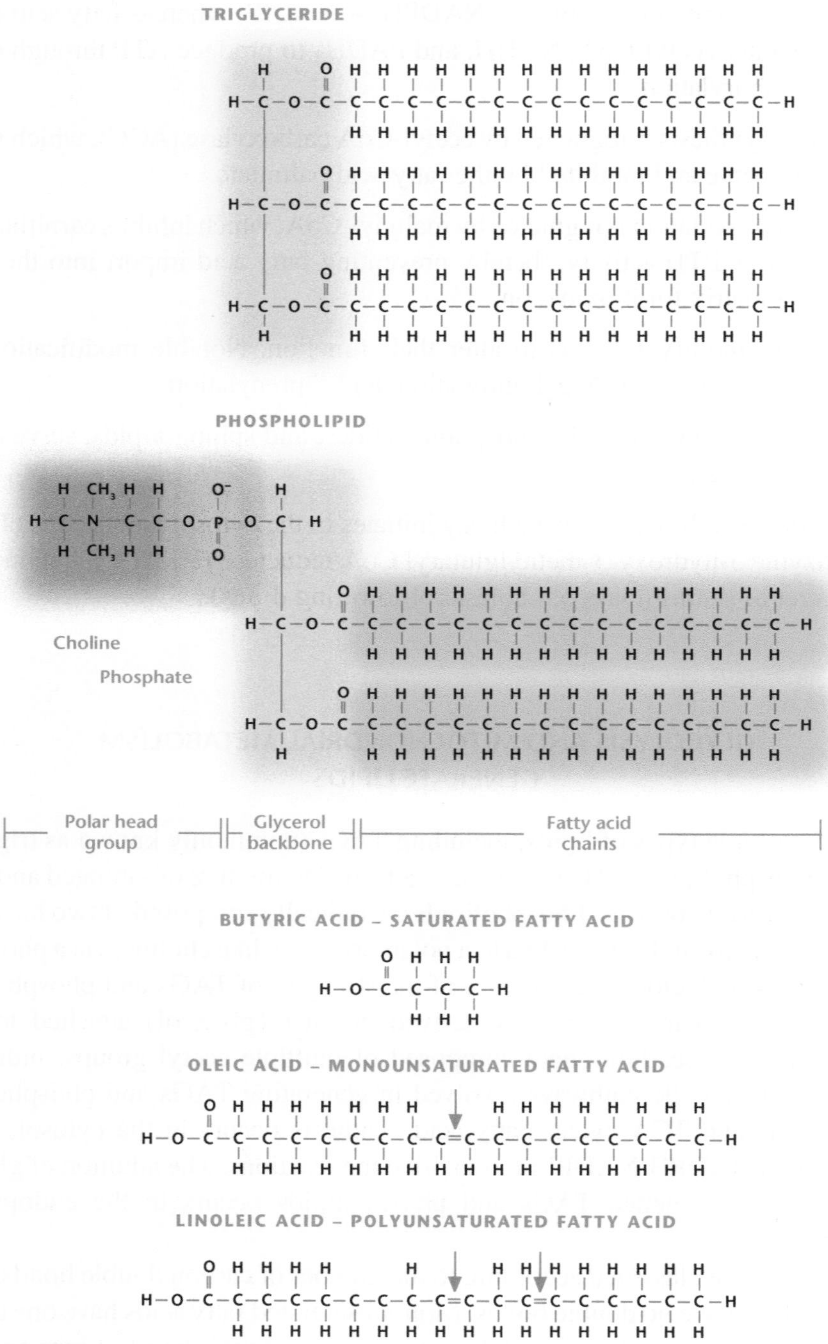

**Figure 7-2.** Structures of lipids. Triglycerides consist of three fatty acids linked to glycerol. Phospholipids contain two fatty acids linked to glycerol that is attached to a polar molecule like choline via phosphate. Saturated fatty acids have no double bonds. Monounsaturated fatty acids have one carbon double bond, whereas polyunsaturated fatty acids have two or more carbon double bonds.

can be modified into other fatty acids by undergoing desaturation to generate unsaturated fatty acids or further chain elongation to produce longer fatty acids.

Acetyl-CoA is the precursor for fatty acid synthesis in the cytoplasm. Acetyl-CoA is generated in the mitochondria from pyruvate (see Chapter 4), which is derived from glucose by glycolysis (see Chapter 3) or amino acid metabolism (see Chapter 8). Acetyl-CoA combines with oxaloacetate and serves as a substrate for citrate synthesis. Acetyl-CoA cannot be transported across mitochondrial membranes; however, the tricarboxylate transporter can transport citrate out of the mitochondria to the cytoplasm. Subsequently, the ATP citrate lyase (ACL) splits citrate into cytoplasmic acetyl-CoA for fatty acid and cholesterol synthesis and oxaloacetate. Malate dehydrogenase reduces oxaloacetate to malate by coupling NADH oxidation to $NAD^+$ (Fig. 7-3). Malate can be transported back into the mitochondrial matrix or undergo oxidation to pyruvate in the cytosol by malic enzyme. The latter generates cytosolic NADPH that can be used to drive fatty acid synthesis. The pyruvate produced returns to the mitochondrial matrix. The pentose phosphate pathway, one-carbon metabolism, and isocitrate dehydrogenase 1 also generate NADPH for fatty acid synthesis (see Chapter 5).

Fatty acid synthesis starts with the irreversible reaction catalyzed by ACC, which causes carboxylation of acetyl-CoA to malonyl-CoA (Fig. 7-4). The Gibbs free energy to drive this reaction comes from $ATP \rightarrow ADP + P_i$. Subsequently, fatty acid synthase, an enzyme encoded by the *FASN* gene, generates palmitate. FASN is a multifunctional protein consisting of two identical multifunctional polypeptides. This complex contains seven different catalytic sites: acetyl transferase, malonyl transferase, β-ketoacyl synthase, β-ketoacyl-acyl carrier protein (ACP) reductase, 3-hydroxyacyl-ACP dehydratase, enoyl-ACP reductase, and thioesterase. These different enzymes are linked covalently in this complex, thus allowing intermediates to be handed efficiently from one active site to another without leaving the assembly.

The intermediates in fatty acid synthesis are linked to the sulfhydryl terminus of a phosphopantetheine group, which is, in turn, attached to a serine residue of the ACP. CoA also contains a phosphopantetheine group. The elongation phase of fatty acid synthesis starts with the formation of acetyl-ACP and malonyl-ACP by acetyl transacylase and malonyl transacylase catalyzing the reactions:

acetyl CoA + ACP $\rightleftharpoons$ acetyl ACP + CoA,          acetyl transacylase,
malonyl CoA + ACP $\rightleftharpoons$ malonyl ACP + CoA,          malonyl transacylase.

Once acetyl-ACP and malonyl-ACP are formed, fatty acids are synthesized by the repetition of the following reaction sequence: condensation $\rightarrow$ reduction $\rightarrow$ dehydration $\rightarrow$ reduction (Fig. 7-4). The acetyl-ACP and malonyl-ACP react to form the four-carbon acetoacetyl-ACP with $CO_2$ as a product by β-ketoacyl-ACP

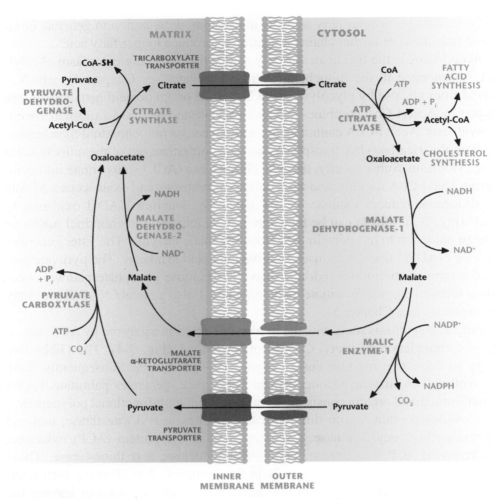

**Figure 7-3.** Mitochondrial citrate generates acetyl-CoA for fatty acid synthesis. Citrate synthase produces citrates from mitochondrial acetyl-CoA and oxaloacetate. Citrate is transported into the cytosol where the ACL splits citrate into cytoplasmic acetyl-CoA and oxaloacetate. Acetyl-CoA is used for fatty acid and cholesterol synthesis. Malate dehydrogenase 1 converts oxaloacetate to malate by coupling NADH oxidation to $NAD^+$. Malate can be transported back into the mitochondrial matrix for regeneration of oxaloacetate or is converted into pyruvate in the cytosol by malic enzyme 1. (Modified, with permission, from Nelson and Cox 2013, p. 841, © W.H. Freeman.)

synthase (also referred to as acyl-malonyl-ACP condensing enzyme). It is important to note that the synthesis of four-carbon acetoacetyl-ACP from two-carbon acetyl-ACP and three-carbon malonyl-ACP reactions is a more favorable reaction than two molecules of acetyl-ACP. ATP is used to carboxylate acetyl-CoA to malonyl-CoA, and the free energy stored in malonyl-CoA is released in the decarboxylation during the generation of acetoacetyl-ACP.

These next three reactions—a reduction, dehydration, and second reduction—convert acetoacetyl-ACP into butyryl-ACP to complete the first elongation cycle.

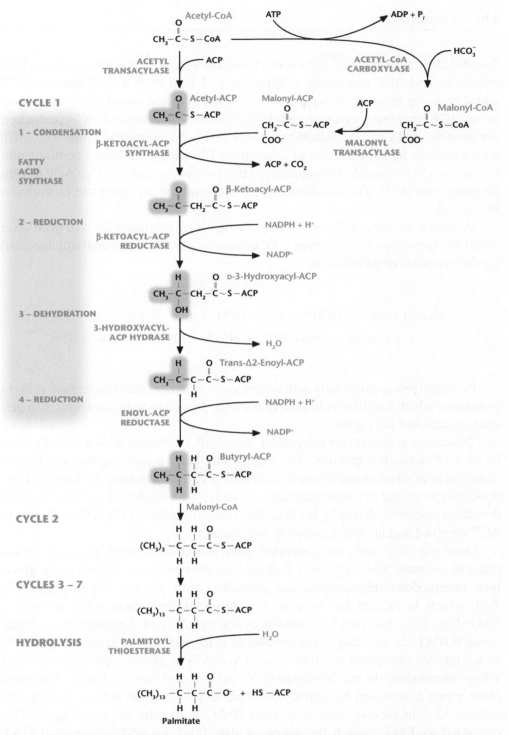

**Figure 7-4.** Fatty acid synthesis pathway. Fatty acid synthesis starts with the irreversible reaction catalyzed by ACC, which carboxylates acetyl-CoA to malonyl-CoA. The elongation phase of fatty acid synthesis starts with the formation of acetyl-ACP and malonyl-ACP from acetyl-CoA and malonyl-CoA by acetyl transacylase and malonyl transacylase, respectively. Fatty acid synthase uses acetyl- and malonyl-ACP to synthesize the 16-carbon fatty acid palmitate by the repetition of the reaction sequence condensation → reduction → dehydration → reduction.

Two NADPH molecules drive these reactions. In cycle 2 of fatty acid synthesis, four-carbon butyryl-ACP condenses with malonyl-ACP to form a $C_6$-β-ketoacyl-ACP, with $CO_2$ being released. This reaction is similar to the first-round reaction, in which acetyl-ACP condenses with malonyl-ACP to form a $C_4$-β-ketoacyl-ACP. Reduction, dehydration, and a second reduction convert the $C_6$-β-ketoacyl-ACP into a $C_6$-ACP, which is ready for a third round of elongation. The elongation cycles continue until $C_{16}$-acyl-ACP is formed. Subsequently, a thioesterase hydrolyzes $C_{16}$-ACP to yield palmitate and ACP. The stoichiometry of the synthesis of palmitate is shown in Figure 7-4.

However, the seven malonyl-CoA molecules were originally derived from seven acetyl-CoA molecules using seven ATP molecules. Hence, the overall stoichiometry for the synthesis of palmitate is

$$8\text{acetyl CoA} + 7\text{ATP} + 14\text{NADPH} + 6\text{H}^+$$
$$\longrightarrow \text{palmitate} + 14\text{NADP}^+ + 8\text{CoA} + 6\text{H}_2\text{O} + 7\text{ADP} + 7\text{P}_i.$$

Palmitate produced by fatty acid synthase can generate other longer fatty acids by elongases, which lengthen palmitate or undergo desaturation by desaturases to generate unsaturated fatty acids.

To activate palmitate for elongation, acyl-CoA synthetase adds a CoA thioester by an ATP-dependent reaction. The elongation occurs by adding malonyl-CoA to palmitate or other saturated or unsaturated fatty acyl-CoA substrates by fatty acyl synthase enzyme on the cytosolic face of the endoplasmic reticulum (Fig. 7-5). This condensation reaction is driven by the decarboxylation of malonyl-CoA. Note there is no ACP involved and no multifunctional enzyme.

Once the fatty acids are generated, they can combine with glycerol 3-phosphate to generate TAG. Glycerol 3-phosphate dehydrogenase converts the glycolytic intermediate dihydroxyacetone phosphate into glycerol 3-phosphate (Fig. 7-5), which is primed for sequential addition of three fatty acids to make a TAG (Fig. 7-5). The first fatty acid is added by glycerol 3-phosphate acyltransferase (GPAT) to generate lysophosphatidic acid, and this is, in turn, acylated by an acylglycerophosphate acyltransferase (AGPAT) to generate phosphatidic acid, a key intermediate in the biosynthesis of all glycerol-derived lipids. The phosphate group is removed by lipins, which act as phosphatidic acid phosphohydrolases (PAPs) to produce diacylglycerols (DAGs). Finally, the resulting DAG is converted to TAG through the action of diacylglycerol acyltransferase (DGAT) enzymes. GPAT, AGPAT, PAPs, and DGATs are localized to the endoplasmic reticulum. The intermediates, phosphatidic acid and DAG, can produce the phospholipids involved in generating membranes, such as cardiolipin and phoshatidyl serine, respectively.

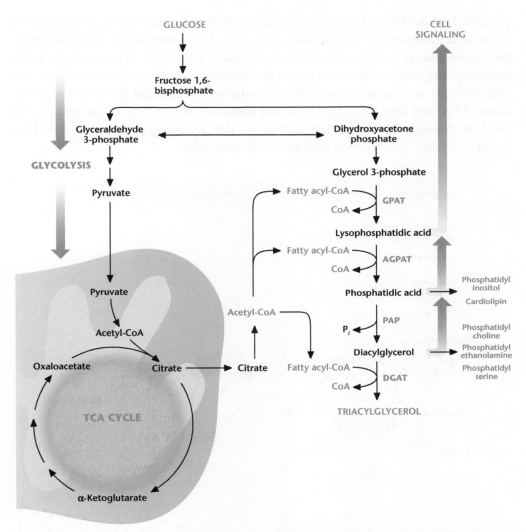

**Figure 7-5.** Synthesis of lipids. Fatty acids combine with glycerol 3-phosphate to generate TAGs and phospholipids through reactions that, in part, take place in the cytosol and endoplasmic reticulum.

## LIPID CATABOLISM GENERATES ATP

TAGs undergo hydrolysis to fatty acid and glycerol by lipases. The glycerol is converted into the glycolysis intermediate dihydroxyacetone phosphate by glycerol kinase and glycerol phosphate dehydrogenase. Fatty acids undergo β-oxidation to generate ATP in the mitochondrial matrix. Fatty acid oxidation is the primary source of ATP when glucose levels are low in cells. If glucose cannot provide pyruvate to generate acetyl-CoA, then fatty acid oxidation provides acetyl-CoA to initiate the TCA cycle. Fatty acids have to be transported from the cytosol to mitochondrial matrix. The first activation step uses fatty acyl-CoA synthetase in the cytosol to generate fatty acyl-CoA in a two-step reaction. In the first step, fatty acid uses ATP to form an acyl-adenylate intermediate (Fig. 7-6). This generates an inorganic pyrophosphate that is rapidly converted to inorganic phosphate. This removal keeps the pyrophosphate concentration low to preserve the favorable reaction. In the second step, the fatty acyl-adenylate intermediate attacked by the thiol group of Co-A generates AMP and fatty acyl-CoA, which is imported into mitochondrial matrix. The AMP is converted by adenylate kinase to ADP by using ATP (AMP + ATP → 2ADP). Overall, this reaction uses two ATP molecules.

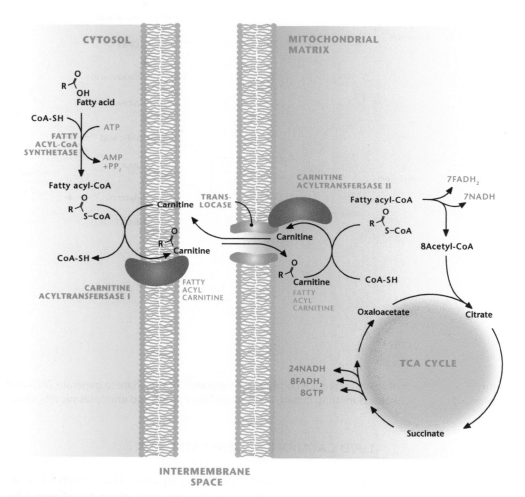

**Figure 7-6.** Carnitine shuttle to transport fatty acids into mitochondria. The import of fatty acyl-CoA into the inner mitochondrial membrane is accomplished by tagging carnitine to fatty acyl-CoA by the carnitine acyltransferase I (CPT1), located in the outer mitochondrial membrane. The carnitine-translocase protein on the inner membrane is an antiporter, which exchanges fatty acyl carnitine for a carnitine. Subsequently, the fatty acyl carnitine in the matrix is converted into fatty acyl-CoA by carnitine acetyltransferase II (CPTII), thereby releasing carnitine, which is shuttled back across the inner membrane to continue the cycle. The number of $FADH_2$ and NADH are shown for palmitate-driven fatty acid oxidation. (Adapted, with permission, from Mehta 2013.)

The import of fatty acyl-CoA into the mitochondrial matrix is accomplished by the carnitine transport cycle (Fig. 7-6). In the first step, carnitine acetyl transferase (CPT1), located in the outer mitochondrial membrane, replaces the CoA moiety with carnitine to form a fatty acyl carnitine molecule, which translocates to the inner mitochondrial membrane. The carnitine-acylcartinine translocase protein, located in the inner mitochondrial membrane, is an antiporter that exchanges fatty acyl carnitine for a carnitine. The fatty acyl carnitine in the matrix is converted into fatty acyl-CoA by carnitine acetyltransferase II, thereby releasing carnitine, which is shuttled back across the inner membrane. Thus, carnitine serves as a tag to get fatty acyl-CoA into the mitochondrial matrix. Subsequently, fatty acyl-CoA enters the β-oxidation pathway, where long-chain fatty acyl-CoA is sequentially degraded into one two-carbon acetyl-CoA, accompanied by the generation of one NADH and one $FADH_2$ by four reactions.

As shown in Figure 7-7, the 16-carbon palmitate fatty acid is converted into palmitoyl-CoA and transferred into the mitochondrial matrix to become the 14-carbon myristoyl-CoA, which is a substrate for another round of β-oxidation. Thus, the complete oxidation of palmitoyl-CoA requires seven rounds of β-oxidation to generate eight molecules of acetyl-CoA, plus seven NADH and $FADH_2$ molecules. The net reaction is

$$palmitoyl\text{-}CoA + 7CoA + 7FAD + 7NAD^+ + 7H_2O$$
$$\longrightarrow 8acetyl\text{-}CoA + 7FADH_2 + 7NADH + 7H^+.$$

Each acetyl-CoA generates 3 NADH, 1 $FADH_2$, and 1 GTP, bringing the total to 24 NADH, 8 $FADH_2$ molecules, and 8 GTP from 8 acetyl-CoA. The total NADH and $FADH_2$ are 31 NADH and 15 $FADH_2$. Remember that each NADH and $FADH_2$ generates ~2.5 ATP and 1.5 ATP through oxidative phosphorylation, respectively (see Chapter 4). This yields 77.5 ATP (31 NADH × 2.5 ATP) plus 22.5 ATP (15 FADH2 × 1.5 ATP) for a total of 100 ATP. The eight GTP can be converted to eight ATP, resulting in 108 ATP. The activation step used two ATP to generate palmitoyl-CoA from palmitate. Thus, after subtracting these two ATP molecules, the final ATP generated by oxidizing palmitate is 106 ATP.

## REGULATION OF FATTY ACID ANABOLISM AND CATABOLISM

Metabolites, hormones, and posttranslational modifications of enzymes all regulate fatty acid synthesis and oxidation. Here, the focus will be on how key intermediates and phosphorylation regulate fatty acid oxidation and synthesis (Fig. 7-8). Fatty acid synthesis is regulated by ACC, citrate, and palmitoyl-CoA. ACC is active in a homopolymeric form and inactive as a monomer. Citrate and palmitoyl-CoA bind to an

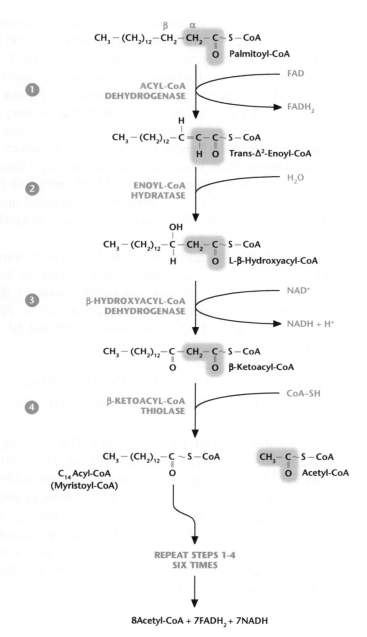

**Figure 7-7.** Mitochondrial β-oxidation. Fatty acids, such as the 16-carbon palmitate fatty acid, are converted into palmitoyl-CoA and transferred into the mitochondrial matrix by the carnitine shuttle. Palmitoyl-CoA enters the β-oxidation pathway and is sequentially degraded into one two-carbon acetyl-CoA accompanied by generation of one NADH and one FADH$_2$ by four reactions to become the 14-carbon myristoyl-CoA, which is a substrate for another round of β-oxidation. The complete oxidation of palmitoyl-CoA requires seven rounds of β-oxidation to generate eight molecules of acetyl-CoA plus seven NADH and FAHD$_2$ molecules.

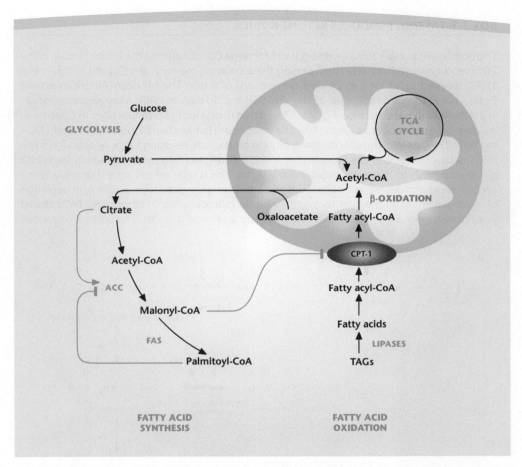

**Figure 7-8.** Metabolic regulation of fatty acid synthesis and oxidation. Acetyl-CoA carboxylase (ACC) is positively and negatively regulated by citrate and palmitoyl-CoA, respectively. An important mechanism to prevent simultaneous fatty acid oxidation and synthesis is an increase in cytosolic malonyl-CoA, which allosterically inhibits CPTI activity to prevent mitochondrial import of fatty acyl-CoA molecules for β-oxidation.

allosteric site on this enzyme to stimulate polymerization or depolymerization, respectively. Acetyl-CoA levels in the cytosol are dependent on how much citrate is exported from mitochondria through the tricarboxylate transporter. As citrate levels accumulate in the cytosol, they activate ACC in a feed-forward mechanism to convert acetyl-CoA to malonyl-CoA, thus, increasing fatty acid synthesis. If the cell accumulates palmitate beyond the cell's metabolic needs, then fatty acid synthesis is diminished. Palmitate can be converted into palmitoyl-CoA to synthesize other fatty acids. The high palmitoyl-CoA levels in the cytosol serve as a feedback inhibitor on ACC and tricarboxylate transporter to decrease flux through the fatty acid synthesis pathway.

## BOX 7-1. FASTING PRODUCES KETONE BODIES

The nonviolent hunger strike campaigns of Mahatma Gandhi against the British Empire in the 20th century are well documented. During these protests, the very lean Gandhi, in 1924 and 1933, survived for as long as 21 days only on sips of water. The Mexican–American activist Cesar Chavez survived fasting for 25 days in 1968 and 36 days in 1988. A key response to fasting is the maintenance of blood glucose levels that is vital for brain to function. In Chapter 6, we covered gluconeogenesis in the liver as a mechanism to maintain blood glucose levels. During starvation, skeletal muscles undergo protein catabolism, resulting in the release of alanine into the blood. The liver absorbs alanine, and it is converted into pyruvate, which feeds into gluconeogenesis to maintain blood glucose levels. This is referred to as the glucose–alanine cycle. Fasting also decreases insulin levels and increases glucagon levels, resulting in depletion of glycogen stores in the liver to maintain blood glucose levels. Furthermore, TAGs stored in adipose tissues are broken down into free fatty acids and glycerol. The latter is converted

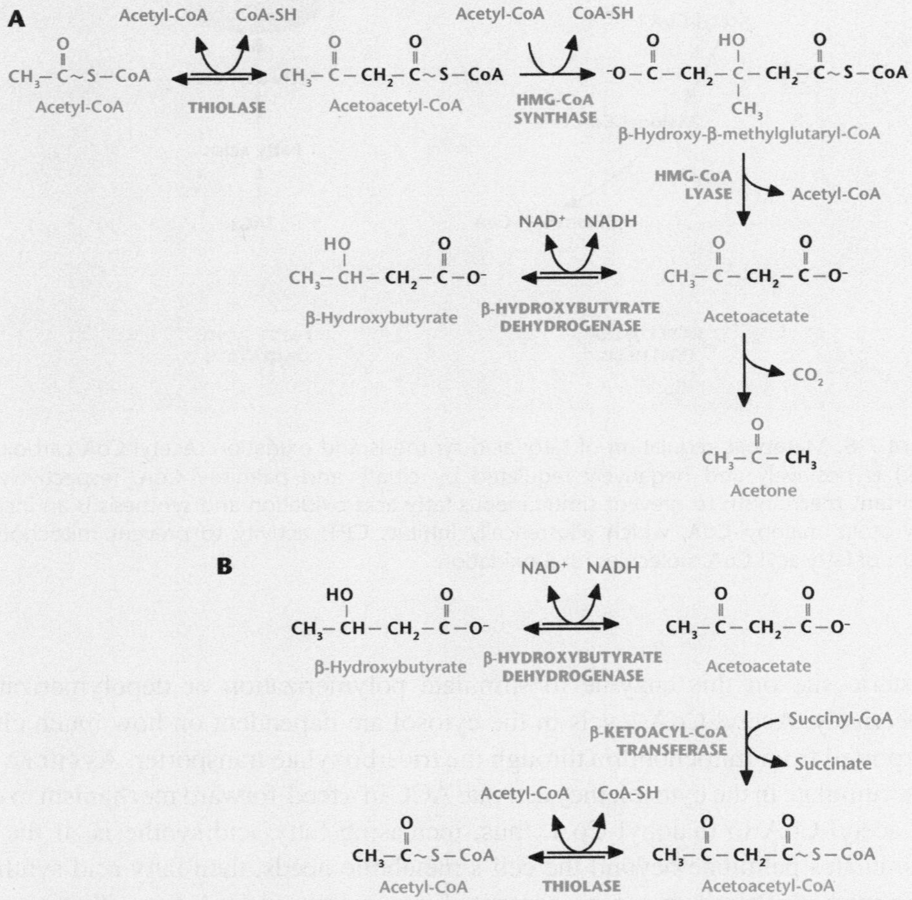

**Box 7-1, Figure 1.** Ketone bodies production and use. Enzymatic reactions that generate the ketone bodies acetoacetate, β-hydroxybutyrate, and acetone. (*B*) Enzymatic reactions that convert ketone bodies into acetyl-CoA.

into glucose by the liver. Free fatty acids can be used by mitochondria to generate ATP through β-oxidation in a variety of energy demanding tissues, such as the heart. The liver uses acetyl-CoA generated from β-oxidation to produce ketone bodies in the form of three molecules: acetoacetate, β-hydroxybutyrate, and acetone (Box 7-1, Fig. 1). The brain does not use free fatty acids, but does use acetoacetate and β-hydroxybutyrate to generate acetyl-CoA for ATP generation. The brain sustains its metabolic rate by using glucose and ketone bodies. The biosynthesis and use of ketone bodies share enzymes with the exception of one enzyme in the ketone body biosynthetic pathway, β-ketoacyl-CoA transferase (Box 7-1, Fig. 1). The liver lacks this enzyme to prevent the futile cycle of synthesis and use of acetoacetate. During starvation, concentrations of ketone bodies increase to millimolar range in blood. People undergoing ketosis (elevated levels of ketone bodies) can be easily detected by the volatile odor of acetone, which is not metabolized for ATP generation. A fascinating new development in the field of ketone bodies is the recognition that ketone bodies are regulators of signaling and dictating biological outcomes. Specifically, β-hydroxybutyrate is an endogenous inhibitor of histone deacetylases.

The other major regulator of fatty acid synthesis is malonyl-CoA that allosterically inhibits CPTI activity to prevent mitochondrial import and degradation of newly synthesized fatty acyl-CoA molecules by β-oxidation. This is an important mechanism to prevent simultaneous fatty acid oxidation and synthesis. An increase in malonyl-CoA serves as a signal to favor fatty acid synthesis over fatty acid oxidation. Conversely, if a cell is in need of ATP, then ACC activity is diminished by phosphorylation by AMP-activated protein kinase (AMPK), leading to a decrease in malonyl-CoA levels and thus relieving the inhibition of carnitine acyltransferase. This promotes fatty acid oxidation and subsequent generation of mitochondrial ATP.

## LIPIDS ACTIVATE CELLULAR SIGNALING PATHWAYS

There are multiple ways that lipids can intersect with signaling, including attachment of lipids to proteins that are necessary for the activity of the protein (Fig. 7-9). Recent studies implicate deregulation of lipid-dependent signaling as an important mechanism for the inflammatory and metabolic diseases. Lipids act as signal transduction messengers at the cell membrane level; a specific lipid can stimulate different cellular responses, depending on cell type and signaling network. The most common function for the attachment of lipids to proteins is to allow water-soluble proteins to interact with hydrophobic membranes. Other functions of lipid modification include forming part of a protein–protein interaction or an integral part of the protein tertiary structure to stabilize the conformation of the protein. It is estimated that 1000 proteins have lipophilic groups covalently attached to them, including fatty acids, phospholipids, sterols, isoprenoids, and glycosylphosphatidyl inositol anchors. All of these modifications confer distinct properties to the modified proteins that are reversible. Notable

---

**BOX 7-2. PEROXISOMES, THE FORGOTTEN ORGANELLES, ALSO CONDUCT FATTY ACID OXIDATION**

Peroxisomes are single membrane-enclosed organelles that are understudied and underappreciated. Mammalian peroxisomes have multiple metabolic functions, including β-oxidation of very-long-chain fatty acids, α-oxidation of branched-chain fatty acids, and synthesis of ether lipids and bile acids, as well as detoxification of ROS. Genetic defects in genes encoding peroxisomal proteins results in a myriad of devastating pathologies. Pertinent to this chapter, let us just focus on oxidation of fatty acids. β-oxidation of very-long-chain fatty acids (more than 26 carbons) ensues primarily in peroxisomes, but not in mitochondria. This oxidation shortens fatty acids that can be further oxidized by mitochondria to generate ATP. Peroxisomes lack a respiratory chain, thus, oxidation of very-long-chain fatty acids to shorter fatty acids generates heat rather than ATP. The transport of fatty acids into peroxisomes also differs from transport into mitochondria. Mitochondria rely on the carnitine exchange system, whereas peroxisomes use three ATP-binding cassette transporter D subfamily proteins that are localized to the peroxisomal membrane: ABCD1, ABCD2, and ABCD3. ABCD1 is mutated in the human disease adrenoleukodystrophy (ALD). The movie *Lorenzo's Oil* is based on the story of a boy who suffered from a deficiency in ABCD1 protein. The inability to oxidize long-chain fatty acids in patients with ALD causes accumulation of these large fatty acids that destroy the myelin sheath "insulation" around nerve cells. The other distinct function of peroxisomes compared with mitochondria is their ability to oxidize branched-chain fatty acid through α-oxidation. Branched-chain fatty acids have a methyl group on the third carbon atom (γ position) that prevents β-oxidation. Therefore, branched-chain fatty acids undergo oxidative decarboxylation (α-oxidation) in peroxisomes to remove the terminal carboxyl group, such as $CO_2$, resulting in a methyl group on the second carbon, which allows for β-oxidation in peroxisomes or mitochondria. The significance of peroxisome's diverse metabolic functions is not fully understood, yet recent studies have linked peroxisome metabolic functions to human pathologies, including cancer, diabetes, and neurodegeneration. Hopefully, the next generation of scientists will embrace this neglected organelle.

---

modifications are *N*-myristoylation, *S*- or *N*-palmitoylation, and *S*-prenylation. Proteins can contain more than one type of these modifications.

*N*-myristoylation is the attachment of a 14-carbon saturated fatty acid, myristate, onto the amino-terminal glycine residue of target proteins and catalyzed by the enzyme *N*-myristoyltransferase. Myristoylated proteins reversibly flip between a myristoyl-accessible state, in which the myristoyl group is available for binding to membranes or other proteins, and a myristoyl-inaccessible state, in which the myristoyl group is located in a hydrophobic-binding pocket within the protein. Myristoylation itself is not strong enough for stable binding to membranes. Thus, many proteins bound to membranes undergo *S*-palmitoylation, which is an attachment of the C16 palmitoyl group from palmitoyl-CoA to the thiolate side chain of cysteine residues within proteins catalyzed by palmitoyl acyltransferases. Because of the longer hydrophobic group compared with the C14 myristoyl moiety, this can permanently anchor the protein to the membrane. Palmitoyl thioesterases cause depalmitoylation of proteins to release proteins from membranes, and this makes

*S*-palmitoylation a reversible switch to regulate membrane localization. Many signaling proteins, such as the Src family of kinases, use *N*-myristoylation and *S*-palmitoylation as a mechanism to localize to the plasma membrane, which is essential for their biological activity (Fig. 7-9).

A second type of palmitate attachment is *N*-palmitoylation, which is essential for the activity of secreted proteins, such as Hedgehog (Hh) and Wnt, that are necessary for proper embryonic development. Hedgehog acyltransferase (Hhat) and porcupine (Porcn) catalyze *N*-palmitoylation of Hh and Wnt proteins in the endoplasmic reticulum lumen to a cysteine and serine residue, respectively. Although cholesterol attachment is not a common lipid modification on proteins, Hh is modified by cholesterol, allowing it to diffuse during development. Many cancer cells also use Hh and Wnt proteins to drive their proliferation and survival. Small-molecule inhibitors against Hhat and Porcn are being tested to block Hh- and Wnt-driven cancers.

Another important lipid modification is *S*-prenylation, which covalently adds a farnesyl (C15) or geranylgeranyl (C20) group to specific cysteine residues within five amino acids from the carboxyl terminus via farnesyl transferase (FT) or geranylgeranyl transferases, respectively. *S*-prenylation of proteins allows them to become membrane-associated because of the hydrophobic nature of farnesyl or geranylgeranyl groups. One example is the Ras GTPase family of proteins, which requires this modification for their optimal activity. Ras can undergo mutations that make it oncogenic, prompting the development of drugs that target FT to inhibit oncogenic Ras activity.

In addition to being posttranslational modifications on proteins, lipids can themselves serve as signaling molecules. Eicosanoids, which are primarily derived from the 20-carbon polyunsaturated fatty acid arachidonic acid, are one example. Eicosa-

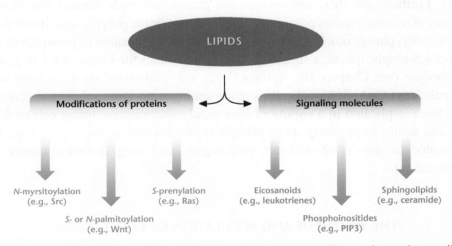

**Figure 7-9.** Overview of lipid signaling. Lipids can modulate signaling pathways by modifying proteins (*N*-myristoylation, *S*- or *N*-palmitoylation, and *S*-prenylation) or serve as signaling molecules (eicosanoids, phosphoinositides, and sphingolipids).

noids are sometimes referred to as local hormones because they are rapidly degraded and have specific effects on targets cells close to their site of production. The three major classes of arachidonate-derived eicosanoids are prostaglandins, leukotrienes, and thromboxanes, which regulate diverse physiological effects, including induction of inflammation, regulation of blood pressure, and blood clotting. The release of eicosanoids affects neighboring cells usually by interacting with their plasma membrane G-protein-coupled receptors to activate diverse cell-signaling pathways.

Eicosanoid synthesis begins with the release of arachidonic acid from phospholipids by phospholipase A2. Arachidonic acid is converted into leukotrienes by lipoxygenase enzymes or into prostaglandin H2 (PGH2) by cyclooxygenase-1 (COX-1) and cyclooxygenase-2 (COX-2) enzymes. PGH2 is a precursor for prostaglandins and thromboxanes. Eicosanoids, derived from PGH2, can amplify inflammation; thus, there are drugs that target COX-1 and COX-2 to reduce inflammation, fever, and pain. These drugs are classified as nonsteroidal anti-inflammatory drugs (NSAIDs). The first NSAID found to inhibit COX-1 and COX-2 was salicylic acid. The German pharmaceutical company Bayer synthesized a pure and stable form of acetylsalicylic acid, which was less irritating than salicylic acid and is marketed as aspirin. Presently, there is host of NSAIDs, such as ibuprofen (Motrin) and naproxen (Alleve), which inhibit COX-1 and COX-2. These molecules can have harmful side effects, such as stomach bleeding. There are multiple other effects exerted by salicylates beyond COX-1 and COX-2 inhibition, including metabolic responses that, in part, are triggered by salicylate activation of AMPK.

Phosphoinositides and sphingolipids are an important class of lipids, worth highlighting, that participate in signaling. Phospholipase C hydrolyzes phosphatidylinositol-4,5-bisphosphate to DAG and inositol-1,4,5-trisphosphate, which trigger the activation of protein kinase C and the release of $Ca^{2+}$ from internal stores, respectively. Furthermore, there are inositol and phosphoinositide kinases that generate an array of soluble inositol polyphosphates and membrane polyphosphoinositide lipids. Notably, phosphoinositide 3-kinase promotes the formation of phosphatidylinositol-3,4,5-trisphosphate, a signaling lipid that activates the kinase AKT to regulate metabolism (see Chapter 10). Sphingolipids are synthesized de novo from serine and palmitoyl-CoA. Sphingolipids, such as sphingosine-1-phosphate and ceramide, have been implicated in a variety of biological outcomes, including cell proliferation and death. Importantly, both phosphoinositides and sphingolipids have been implicated as drivers of multiple pathologies, including diabetes, cancer, and inflammation.

## THE SYNTHESIS AND REGULATION OF CHOLESTEROL

Cholesterol becomes an obsession for many as they age. You cannot help but notice how many people having dinner at a restaurant or party will obsess about whether their cholesterol levels are high or if they should take cholesterol-lowering drugs. Although high cholesterol levels need to be dealt with, it is important to realize that cholesterol has important biological functions. Cholesterol can incorporate into cell membranes, function as a modification on proteins, and serve as precursor for the generation of steroid hormones and vitamin D. Cholesterol is not essential in our diet because most cells can generate cholesterol. The highest producer of cholesterol is the liver, where it can be esterified into cholesterol esters, which are stored in lipid droplets, packaged into lipoprotein particles, and exported to the peripheral tissues, or converted into bile acids, which are secreted into the small intestine through the bile duct to act as emulsifying agents in the digestion of dietary fat. Some of the bile is excreted as waste, thus getting rid of excess cholesterol, but most returns to the liver and then is moved to the gall bladder for storage and use in the digestion of lipids. The most abundant bile acid is cholic acid (cholate).

The cholesterol biosynthetic pathway takes place in the cytosol and is initiated by the enzyme HMG-CoA reductase, the target of statin drugs (see Box 7-3), to generate the six-carbon molecule mevalonate from one molecule of acetyl-CoA and one molecule of acetoacetyl-CoA (Fig. 7-10). Mevalonate is then phosphorylated and decarboxylated to form the activated 5-carbon isoprenoid intermediate, isopentenyl pyrophosphate. Next, three molecules of isopentenyl pyrophosphate are combined to form farnesyl pyrophosphate (C15), which is then used to generate squalene, a C30 cholesterol precursor. C15 can be used to prenylate proteins, such as Ras. Squalene synthase catalyzes condensation of two molecules of C15 with reduction by NADPH to generate squalene, which undergoes multiple reactions to generate a 27-carbon cholesterol molecule.

Important by-products derived from cholesterol are the steroid hormones. Most of us are familiar with the steroid hormones estrogen and testosterone because of their pivotal role in reproductive physiology in females and males, respectively. These two hormones, along with the glucocorticoid cortisol and mineralocorticoid aldosterone, are derived from progesterone (Fig. 7-10). Cortisol induces gluconeogenesis to increase blood sugar and is a powerful suppressor of the immune system; aldosterone regulates ion transport in the kidney. Cholesterol generates pregnenolone to produce progesterone. Because of their pleiotropic roles, synthetic steroids have been generated as pharmacological agents to treat a wide range of pathologies, including asthma (glucocorticoids).

## MULTIPLE MECHANISMS REGULATE CELLULAR CHOLESTEROL LEVELS

Cholesterol is essential for multiple biological processes, thus cells have mechanisms for short- and long-term regulation of cholesterol synthesis. The rate-limiting step in cholesterol synthesis is HMG-CoA reductase, which localized to the endoplasmic reticulum membrane. AMPK, which phosphorylates to inhibit the activity of this

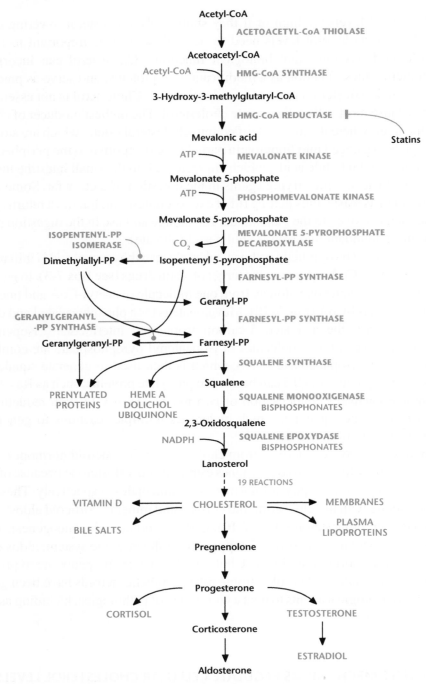

**Figure 7-10.** Cholesterol biosynthesis pathway. The cholesterol biosynthetic pathway is initiated by the enzyme HMG-CoA reductase, the target of statin drugs, to generate the six-carbon molecule mevalonate from one molecule of acetyl-CoA and one molecule of acetoacetyl-CoA. Mevalonate undergoes multiple reactions to generate a 27-carbon cholesterol molecule. Important by-products derived from cholesterol are the steroid hormones, bile salts, vitamin D, plasma lipoproteins, and structural elements of membranes.

enzyme, regulates this enzyme in metabolically stressed cells to shut off cholesterol synthesis. As we have already discussed, AMPK activation also stimulates fatty acid oxidation to stimulate mitochondrial-generated ATP. Remember, metabolically stressed cells will typically shut down ATP-consuming anabolic functions and concomitantly promote ATP-generating catabolic functions to maintain a high ATP/ADP ratio.

Long-term regulation of cholesterol synthesis is accomplished by sensing fluctuations in levels of cholesterol and other sterols in the pathway, triggering changes in multiple genes in the cholesterol pathway, including HMG-CoA reductase. Cholesterol is embedded in membranes, and, therefore, sensing changes in sterol levels occurs in membranes by two proteins embedded in endoplasmic reticulum membranes that contain sterol-sensing domains, HMG-CoA and SREBP cleavage-activating protein (SCAP). HMG-CoA detects increases in sterols through its sterol-sensing domain, resulting in binding to two other ER membrane proteins called INSIG-1 and INSIG-2 (insulin-induced gene 1 and 2), which are bound to ubiquitin ligases. This triggers proteasomal degradation of HMG-CoA protein.

SCAP is essential for the activation of a family of transcription factors SREBP (sterol regulatory element-binding protein) that control genes involved in cholesterol and fatty acid synthesis. There are three closely related isoforms of SREBP in mammalian cells, referred to as SREBP1a, SREBP1c, and SREBP2. SREBP1a and SREBP1c are generated from the same gene through alternative splicing and use of different promoters and are involved in fatty acid synthesis. SREBP1c and SREBP2 control genes involved in fatty acid and cholesterol synthesis, respectively. SREBP1a can control genes in both pathways. SREBPs belong to the basic helix-loop-helix–leucine zipper (bHLH-Zip) family of transcription factors. However, they are different from other bHLH-Zip proteins in that they are synthesized as inactive precursors bound to the endoplasmic reticulum; to be transcriptionally active, they have to reach the nucleus. When cholesterol levels become low, SCAP binds to SREBPs and escorts them from the endoplasmic reticulum to the Golgi apparatus. Next, SREBPs undergo sequential proteolytic processing by Site-1 and Site-2 protease to release the amino-terminal bHLH-Zip domain from the Golgi membrane. The bHLH-Zip domain enters the nucleus and binds to a sterol response element in the enhancer/promoter region of target genes to activate their transcription. When cholesterol increases, SCAP binds to INSIGs, trapping the SREBP complex in the endoplasmic reticulum membrane. SREBPs are not able to reach the Golgi apparatus and the bHLH-Zip domain is not released to activate gene transcription.

It is important to note that two mechanisms of cholesterol pathway regulation, HMG-CoA degradation and SCAP regulation of SREBP, respond to different sterols. Lanosterol primarily triggers degradation of HMG-CoA and cholesterol inhibits SREBP activation. Lanosterol, the cholesterol precursor, is a more potent inducer of HMG-CoA degradation compared with cholesterol. Lanosterol is known to be toxic to cells; it makes sense to have it degrade HMG-CoA protein as its levels increase to shut off the pathway. If lanosterol accumulation were to shut off SREBP processing, rather than degrading HMG-CoA, then it would decrease the enzymes required for lanosterol conversion to cholesterol and this could result in transient increase in lanosterol. Thus, having lanosterol degrade HMG-CoA reductase, and not SREBP, shuts off further synthesis of lanosterol while allowing for the enzymes to convert lanosterol to cholesterol. As cholesterol increases, the SREBP genes are turned off to shut down this pathway.

The elucidation of the cholesterol pathway led to the mechanism by which statins decrease cholesterol levels in the blood, and, at first glance, this mechanism would reduce cholesterol synthesis by HMG-CoA reductase inhibition. However, the mechanism is a bit more complicated and involves induction of the low-density lipoprotein (LDL) receptor, discovered by Michael Brown and Joseph Goldstein in the 1970s. They set out to elucidate the underlying mechanism of a human disease called familial hypercholesterolemia (FH). The concentration of cholesterol in blood of patients with FH is abnormally high, and these individuals are at high risk of a heart attack early in life. By examining fibroblasts isolated from FH homozygous

---

**BOX 7-3. THE STORY OF STATINS AS CHOLESTEROL-REDUCING AGENTS**

A detrimental effect of rising cholesterol levels in the blood is that they are associated with the formation of atherosclerotic plaques in the lining of blood vessels. If atherosclerotic plaques rupture, pieces of fibrous material break off and travel to smaller blood vessels, where they cause blood clots (thrombosis) that can block blood flow to vital organs, such as the brain and heart, causing stroke and heart attack. In thinking about how to lower cholesterol levels to reduce atherosclerotic plaques, scientists targeted the cholesterol biosynthesis pathway. Initial efforts targeted late steps in the pathway using the compound Triparanol. However, this drug was withdrawn from clinical use because of adverse effects, including development of cataracts. Triparanol caused the accumulation of desmosterol by inhibiting the enzyme that converts desmosterol to cholesterol. In contrast, inhibition of HMG-CoA reductase, early in the cholesterol pathway, results in accumulation of hydromethylglutrate, which is not toxic and has alternative metabolic pathways for its breakdown. In the 1970s, the Japanese microbiologist Akira Endo discovered ML236B (compactin or mevastatin) in a fermentation broth of *Penicillium citrinum* as an HMG-CoA reductase inhibitor. Sankyo, in Japan, developed this compound for clinical use; however, this inhibitor never made it to the market because of fears of toxicity. The inhibitor that eventually made it to the market was lovastatin in 1987. This inhibitor, initially named mevinolin, was discovered at Merck Research Laboratories in a fermentation broth of *Aspergillus terreus* in the late 1970s. Subsequently, the second statin approved was simvastatin (Zocor) in 1988, which differs from lovastatin by having an additional methyl group as a side chain. Pravastatin (Pravachol), derived from compactin by Sankyo, followed in 1991. The next statins, fluvastatin (Lescolin) in 1994, atorvastatin (Lipitor) in 1997, cerivastatin (Baycol) in 1998, and rosuvastatin (Crestor) in 2003, were all synthetic compounds and not derived from microorganisms. For his pioneering efforts in discovering the original statin compactin, Dr. Endo received the prestigious 2008 Lasker-DeBakey Award for Clinical Medical Research. Forty years after his initial discovery, millions have used statins to reduce the morbidity and mortality associated with cardiovascular diseases.

patients, they were able to reveal that these cells lack their ability to take up LDL. Subsequently, experiments showed that LDL binding to the LDL receptor at specific sites on the membranes, called clathrin-coated pits, triggered internalization of the receptor, followed by lysosomal hydrolysis to free the cholesterol from the LDL particle. The increase in cholesterol inhibited HMG-CoA reductase activity by degradation of the protein and decrease in transcription through inhibiting SREBP, as discussed earlier. SREBP also induces the LDL receptor. If SREBP is inhibited by the increase in cholesterol because of internalization of the LDL receptor, transcription of the LDL receptors decreases. An increase in cholesterol levels also activates the cholesterol esterfying enzyme, cholesterol acyl-transferase, which allows cholesterol to be stored in cells as ester droplets. Thus, when statins directly inhibit HMG-CoA reductase, this triggers a decrease in cholesterol, which activates SREBP-mediated induction of the LDL receptors on the surface of cells and subsequent endocytosis of LDL receptor particles from the blood. The particles taken up by the liver are shunted off to bile acids for storage and excretion from the body. Statins reduce LDL levels in serum through this mechanism by 20%–40%. This discovery of Brown and Goldstein resulted in their being awarded the 1985 Nobel Prize in Physiology or Medicine "for their discoveries concerning the regulation of cholesterol metabolism." Beyond cholesterol metabolism, the importance of their discovery lies in concept of receptor-mediated endocytosis and receptor recycling, which provided a conceptual framework by which cells can internalize hormones, growth factors, and viruses. The story of cholesterol biology and statins is a good example of how a mechanistic reductionist approach can elucidate fundamental biology and provide benefit to patients.

## REFERENCES

Mehta SA. 2013. Activation and transportation of fatty acids to the mitochondria via the carnitine shuttle with activation. http://pharmaxchange.info/press/2013/10/activation-and-transportation-of-fatty-acids-to-the-mitochondria-via-the-carnitine-shuttle-with-animation/.

Nelson DL, Cox MM. 2013. *Lehninger principles of biochemistry*, 6th ed. WH Freeman, New York.

## ADDITIONAL READING

Foster DW. 2012. Malonyl-CoA: The regulator of fatty acid synthesis and oxidation. *J Clin Invest* **122:** 1958–1959.

Goldstein JL, Brown MS. 2009. The LDL receptor. *Arterioscler Thromb Vasc Biol* **29:** 431–438.

Goldstein JL, DeBose-Boyd RA, Brown MS. 2006. Protein sensors for membrane sterols. *Cell* **124:** 35–46.

Lodhi IJ, Semenkovich CF. 2014. Peroxisomes: A nexus for lipid metabolism and cellular signaling. *Cell Metab* **19:** 380–392.

Menendez JA, Lupu R. 2007. Fatty acid synthase and the lipogenic phenotype in cancer pathogenesis. *Nat Rev Cancer* **7:** 763–777.

Newman JC, Verdin E. 2014. Ketone bodies as signaling metabolites. *Trends Endocrinol Metab* **25:** 42–52.

Resh MD. 2013. Covalent lipid modifications of proteins. *Curr Biol* **23:** R431–R435.

Tobert JA. 2003. Lovastatin and beyond: The history of the HMG-CoA reductase inhibitors. *Nat Rev Drug Discov* **2:** 517–526.

Wymann MP, Schneiter R. 2008. Lipid signalling in disease. *Nat Rev Mol Cell Biol* **9:** 162–176.

# 8

## Amino Acids

THE LOCAL VITAMIN AND NUTRITION STORE IS popular in my neighborhood. Recently, I walked into this store and realized it is similar to the biotechnology companies in which many scientists purchase their amino acids, media, and cell-culture reagents. There were the usual supplements to treat every disease you can imagine, but by far the most popular items showcased were amino acids. The clerk told me that many people purchase amino acids and spike their smoothie drinks to boost muscle performance and enhancement. Of course, many of us prefer to eat eggs, lentils, fish, and meat as a source of protein, which is digested to generate the amino acids required for intracellular de novo protein synthesis. There are 20 amino acids. Animals have seven conditionally essential amino acids that can be synthesized and are usually not required in the diet. However, they are essential components of the diet for specific populations that cannot synthesize them in adequate amounts. There are four "nonessential" amino acids can be synthesized and, thus, are not required in diet. The remaining nine "essential" amino acids are obtained from diet. Most plants and bacteria, however, can synthesize all 20 amino acids. Most of us usually think of amino acids as important for protein synthesis, but they can also generate glucose, ATP, and fatty acids, and they are metabolic precursors for numerous biomolecules, including heme groups, nucleotide bases, and signaling molecules (catecholamines, neurotransmitters). In addition, they are essential for epigenetic modifications. This chapter covers (1) generation of amino acids, (2) degradation of amino acids, and (3) generation of catecholamines and heme by amino acids and their role in epigenetics (Fig. 8-1).

### QUICK GUIDE TO AMINO ACIDS

- There are 20 amino acids. Animals cannot synthesize nine amino acids, which are called "essential" amino acids; therefore, they are taken from diet (Fig. 8-1). The remaining 11 are "nonessential" or "conditionally essential" amino acids.
- Glycolytic and TCA cycle intermediates generate the nonessential and conditionally essential amino acids.

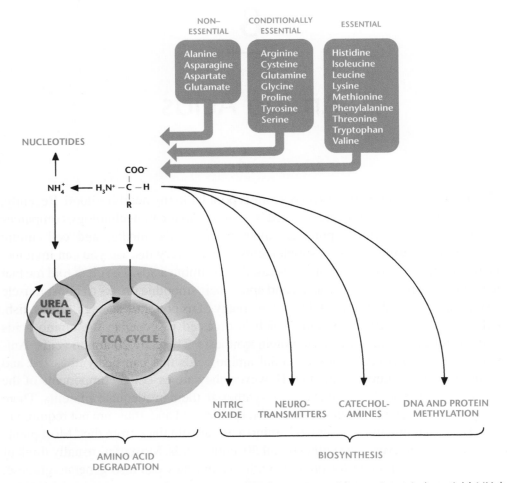

**Figure 8-1.** An overview of amino acid metabolism. Amino acids are degraded to yield $NH_4^+$, which enters the urea cycle, and a carbon skeleton that can enter metabolic pathways to generate ATP, glucose, and fatty acids. Amino acids are also used to produce NO, neurotransmitters, and catecholamines. The amino acid methionine provides the methyl group for many DNA and histone methyltransferases to regulate epigenetics.

- Free amino acids produced from degradation of either cellular proteins or dietary proteins are deaminated to yield $NH_4^+$ and a carbon skeleton. $NH_4^+$ enters the urea cycle and the carbon skeleton can enter metabolic pathways to generate ATP, glucose, and fatty acids (Fig. 8-1).

- Glutamate acts as both a nitrogen donor and acceptor and is the central amino acid for the movement of nitrogen among amino acids.

- Tyrosine is the precursor to generation of norepinephrine, epinephrine, dopamine, and melanins.

- Methionine provides the methyl group for many DNA and histone methyltransferases to regulate epigenetics.

- Cysteine, glutamate, and glycine generate the antioxidant glutathione.

- Nitric oxide synthases use arginine to generate nitric oxide (NO).

- Glycine and glutamate can serve as neurotransmitters. Glutamate can generate another amino acid neurotransmitter, γ-aminobutyric acid (GABA), which does not participate in protein synthesis.

- Tryptophan is used as a precursor for generation of another neurotransmitter, serotonin. Serotonin is used to generate melatonin.

### METABOLIC PATHWAYS THAT PRODUCE NONESSENTIAL AMINO ACIDS

Animals lack the enzymes to generate the essential amino acids, thus, these amino acids must be obtained from the diet. The structures of the essential amino acids, generally, are more complex than the nonessential amino acids. Essential amino acids require a significantly greater number of enzymatic reactions for synthesis and are found in plants and lower organisms. Tyrosine and arginine are conditionally essential. Tyrosine is derived from the essential amino acid phenylalanine by the enzyme phenylalanine hydroxylase. Endogenous tyrosine production is dependent on dietary phenylalanine, and a significant portion of tyrosine is obtained from diet. A small amount of arginine can be generated from argininosuccinate in the urea cycle (see the section The Urea Cycle Is Necessary for Amino Acid Degradation).

Animals can synthesize the conditionally essential and nonessential amino acids using glycolytic and TCA cycle intermediates. The glycolytic intermediate 3-phosphoglycerate generates serine and glycine. These reactions are outlined in Figure 8-2. In the initial step, 3-phosphoglycerate is oxidized by 3-phosphoglycerate dehydrogenase to generate 3-phosphohydroxypyruvate by coupling this reaction to the reduction of NAD$^+$ to NADH. Subsequently, 3-phosphohydroxypyruvate undergoes a glutamate-linked transamination reaction catalyzed by phosphoserine aminotransferase 1 to form 3-phosphoserine, which is then hydrolyzed by phosphoserine phosphatase to yield serine. The enzyme serine hydroxytmethyltransferase converts serine into glycine, coupling this reaction to the conversion of tetrahydrofolate (THF) to $N^5,N^{10}$-methylene THF. Serine is also required for cysteine production. Cystathionine β synthase converts homocysteine and serine into cystathionine, which cystathionine γ-lyase converts to cysteine (see Fig. 8-11).

An important role of glycine is that it is a metabolic precursor to heme, as shown in Figure 8-3. The initial step catalyzed by δ-aminolevulinic acid synthase combines glycine with the TCA cycle intermediate succinyl-CoA to generate δ-aminolevulinic acid in the mitochondrial matrix. Next, δ-aminolevulinic acid is exported to the cytosol, where two molecules condense to generate porphobilinogen. A series of reactions occur in the cytosol and mitochondria, culminating in the final step in which ferrochelatase incorporates Fe$^{2+}$ into the heme ring.

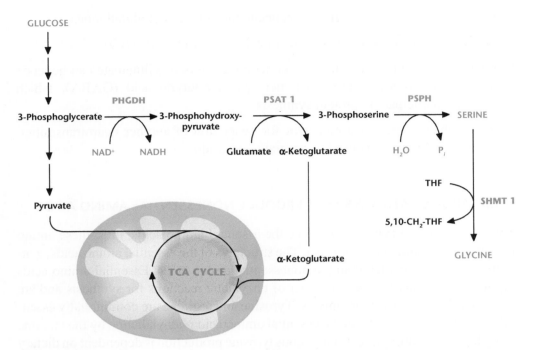

**Figure 8-2.** The glycolytic intermediate 3-phosphoglycerate generates serine and glycine. 3-Phosphoglycerate dehydrogenase (PHGDH) oxidizes 3-phosphoglycerate to generate 3-phospho-hydroxypyruvate, which is converted into 3-phosphoserine by phosphoserine aminotransferase 1 (PSAT1). Subsequently, phosphoserine phosphatase (PSPH) converts 3-phosphoserine into serine. The enzyme serine hydroxytmethyltransferase (SHMT1) converts serine into glycine by coupling this reaction to the conversion of THF to $N^5,N^{10}$-methylenetetrahydrofolate (5,10-CH2-MTHF).

The major TCA cycle intermediate α-ketoglutarate is at the heart of generating multiple amino acids. In particular, the interconversion of glutamate to α-ketoglutarate allows the generation of multiple amino acids, including alanine, aspartate, and arginine. Glutamate undergoes two transamination reactions, which generate alanine and aspartate, both in the cytosol and mitochondrial matrix. Aspartate aminotransferase uses glutamate and oxaloacetate to generate aspartate and α-ketoglutarate, and alanine aminotransferase uses glutamate and pyruvate to generate alanine and α-ketoglutarate (Fig. 8-4). Ornithine aminotransferase generates ornithine by coupling to glutamate conversion into α-ketoglutarate. Subsequently, ornithine can generate arginine through the urea cycle (see Fig. 8-7). Glutamate can be generated from α-ketoglutarate and glutamine. Glutamate dehydrogenase converts α-ketoglutarate and $NH_4^+$ to glutamate by coupling NADPH to $NADP^+$. Glutamine synthetase uses ATP and $NH_4^+$ to generate glutamine from glutamate. Conversely, glutaminase converts glutamine into glutamate with the release of ammonium ($NH_4^+$). Glutamine can also be converted into glutamate and asparagine by asparagine synthetase, a reaction that uses aspartate and ATP as substrates (Fig. 8-4). Finally, γ-glutamyl kinase uses ATP to convert glutamate into glutamate 5-phosphate, which is then converted

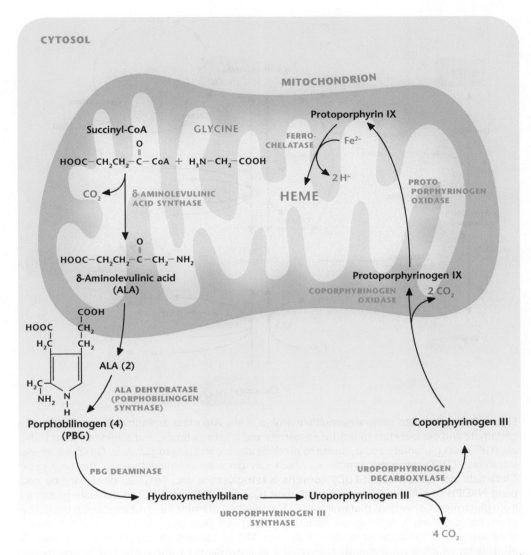

**Figure 8-3.** Glycine is a precursor in heme synthesis. The initial step of heme synthesis is catalyzed by δ-aminolevulinic acid synthase, which combines glycine with the TCA cycle intermediate succinyl-CoA to generate δ-aminolevulinic acid in the mitochondrial matrix. Next, δ-aminolevulinic undergoes a series of reactions, which occur in the cytosol and mitochondria, to generate the heme ring.

into δ-pyrroline-5-carboxylate and proline by pyrroline-5-carboxylate reductase (Fig. 8-4).

## THE UREA CYCLE IS NECESSARY FOR AMINO ACID DEGRADATION

Free amino acids are generated from degradation of either cellular proteins or dietary proteins. Free amino acids cannot be simply stored; thus, they are recycled for protein synthesis or deaminated to yield $NH_4^+$ and a carbon skeleton. The $NH_4^+$ is toxic to

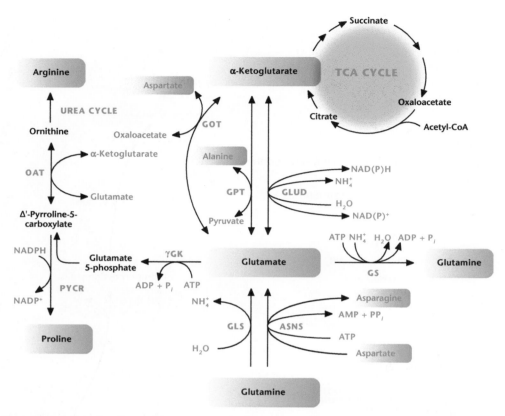

**Figure 8-4.** Glutamate generates multiple amino acids. Aspartate aminotransferase (GOT) uses glutamate and oxaloacetate to produce aspartate and α-ketoglutarate, and alanine aminotransferase (GPT) uses glutamate and pyruvate to produce alanine and α-ketoglutarate. Ornithine aminotransferase (OAT) generates ornithine, which can generate arginine through the urea cycle. Glutamate dehydrogenase (GLUD) converts α-ketoglutarate and $NH_4^+$ to glutamate by coupling NADPH to $NADP^+$. Glutamine synthetase (GS) uses ATP and $NH_4^+$ to generate glutamine from glutamate. Conversely, glutaminase (GLS) converts glutamine into glutamate with the release of ammonium ($NH_4^+$). Asparagine synthetase (ASNS) uses aspartate, glutamine, and ATP to generate asparagine. γ-Glutamyl kinase (γGK) uses ATP to convert glutamate into glutamate 5-phosphate, which is subsequently converted into δ-pyrroline-5-carboxylate and proline by pyrroline-5-carboxylate reductase (PYCR).

cells and removed either by synthesis of nitrogen-containing compounds, such as nucleotides or amino acids, or excreted in the form of urea. The carbon skeletons of amino acids feed into metabolic pathways to generate ATP, glucose, and fatty acids (Fig. 8-5).

The carbon skeleton of amino acids can be converted into TCA cycle intermediates that can be used either to generate ATP through oxidative phosphorylation or provide the precursors for fatty acid synthesis and gluconeogenesis (Fig. 8-5). Amino acids normally account for 10%–15% of ATP generated. However, under high-protein diets or conditions of starvation in which the muscle protein is

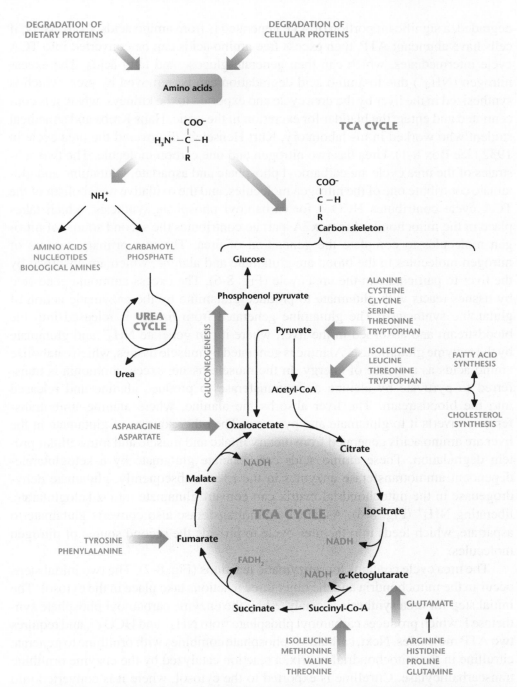

**Figure 8-5.** Amino acid degradation produces ammonium ($NH_4^+$) and a carbon skeleton. The $NH_4^+$ is removed either by synthesis of nitrogen-containing compounds, such as nucleotides, or excreted in the form of urea. The carbon skeletons of amino acids can be converted into TCA cycle intermediates, which are used either to generate ATP through oxidative phosphorylation or provide the precursors for fatty acid synthesis and gluconeogenesis.

degraded, a significant portion of ATP generated is from amino acids. Conversely, if cells have abundant ATP, then excess free amino acids can be converted into TCA cycle intermediates, which can then generate glucose and fatty acids. The excess nitrogen ($NH_4^+$) due to amino acid degradation can be removed by urea, which is synthesized in the liver by the urea cycle and exported to the kidneys, where it is concentrated and enters the bladder for excretion in the urine. Hans Krebs and a medical student who worked in his laboratory, Kurt Henseleit, discovered the urea cycle in 1932 (see Box 8-1). Urea has two nitrogen and one carbon molecule. The two substrates of the urea cycle are carbamoyl phosphate and aspartate. Glutamine and glutamate contribute one of the nitrogen molecules, and the oxidative metabolism of the TCA cycle contributes $HCO_3^-$ for carbamoyl phosphate synthesis, which takes place in the mitochondrial matrix. Aspartate contributes the second source of nitrogen molecules to complete the generation of urea. The two primary carriers of nitrogen molecules in the blood are glutamine and alanine, which are absorbed by the liver to participate in the urea cycle (Fig. 8-6). The excess ammonia generated by tissues reacts with glutamate to produce glutamine by the enzymatic action of glutamine synthetase. The glutamine generated from tissues is released into the bloodstream and absorbed in the liver, where it can generate $NH_4^+$ and glutamate by the enzyme glutaminase. Alanine is generated by muscle tissues, which catabolize amino acids as a source of energy. In the muscle tissue, excess ammonia is transferred to pyruvate by alanine aminotransferase to produce alanine and released into the bloodstream. The liver absorbs the alanine, where alanine-aminotransferase converts it to glutamate and pyruvate. The other sources of glutamate in the liver are amino acids generated from dietary intake and muscle and intracellular protein degradation. These amino acids can generate glutamate by $\alpha$-ketoglutarate-dependent aminotransferase enzymes in the liver. Subsequently, glutamate dehydrogenase in the mitochondrial matrix can convert glutamate into $\alpha$-ketoglutarate, liberating $NH_4^+$ (Fig. 8-6). Aspartate aminotransferase also converts glutamate to aspartate, which feeds into the urea cycle to provide the second source of nitrogen molecules.

The urea cycle contains five enzymatic reactions (Fig. 8-7). The two initial steps occur in the mitochondria and the other three reactions take place in the cytosol. The initial step for urea synthesis is catalyzed by the enzyme carbamoyl phosphate synthetase I, which produces carbamoyl phosphate from $NH_4^+$ and $HCO_3^-$, and requires two ATP molecules. Next, carbamoyl phosphate combines with ornithine to generate citrulline in the mitochondrial matrix, a reaction catalyzed by the enzyme ornithine transcarbamoylase. Citrulline is exported to the cytosol, where it is converted into argininosuccinate by argininosuccinate synthetase. This reaction results in the incorporation of a second nitrogen atom from aspartate. In the next reaction, the enzyme argininosuccinase cleaves argininosuccinate to produce fumarate and arginine. Finally, the enzyme arginase converts arginine to urea and ornithine to complete the urea cycle. Note that ornithine is recycled in a fashion similar to oxaloacetate

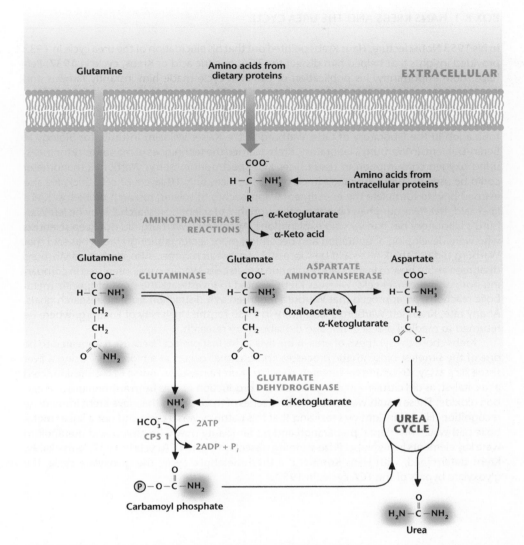

**Figure 8-6.** Glutamine and glutamate generate two nitrogen molecules for the urea cycle. Urea contains two nitrogen molecules. Glutaminase generates $NH_4^+$ and glutamate. Subsequently, glutamate can be converted into $\alpha$-ketoglutarate to liberate another $NH_4^+$ by glutamate dehydrogenase. Carbamoyl phosphate synthetase (CPS1) uses $NH_4^+$, $HCO_3^-$, and ATP as substrates to generate carbamoyl phosphate, which provides the first source of nitrogen molecules for urea generation. Amino acids, such as alanine, can also be converted into glutamate through aminotransferases. Next, the aspartate aminotransferase converts glutamate into aspartate, which feeds into the urea cycle to provide the second source of nitrogen molecules for urea production.

## BOX 8-1. HANS KREBS AND THE UREA CYCLE

In his 1953 Nobel lecture, Hans Krebs pointed out that his elucidation of the urea cycle in 1932 provided insights that helped him dissect the TCA, or citric acid or Krebs, cycle in 1937. Perhaps more importantly, his publication of the urea cycle made him instantly famous and was instrumental in enabling Krebs, a Jew, to escape Nazi Germany.

Krebs was born in Germany in 1900 and was initially destined to become an otolaryngologist like his father. But after receiving his M.D. from the University of Hamburg in 1925, he took a job in the laboratory of Otto Warburg at the Kaiser Wilhelm Institute for Biology at Berlin-Dahlem. In Warburg's laboratory, Krebs learned the techniques of manometry for measuring oxygen consumption in tissue slices and spectrophotometry. Warburg's manometer could be used to study respiration in vitro in tissue slices only 10-layers-of-cells thick. He also learned how to formulate the experimental approaches to solving research problems. Krebs later said that "Warburg had taught me the alphabet of scientific research." Why he left Warburg's laboratory has had varying explanations. Apparently, Warburg did not keep scientists who were developing a reputation and becoming more senior, although Krebs has said that Warburg did not think he would be successful in a research career. Also, Krebs and Warburg disagreed about how to investigate respiration within cells. Warburg was interested in comparing normal and cancer cells, whereas Krebs wanted to investigate the steps of specific metabolic reactions, an approach that Warburg regarded as a distraction from his research goals. At any rate, Krebs left Warburg's laboratory in 1930 for the University of Freiburg, where he returned to medical practice, but also did laboratory research.

Krebs chose the synthesis of urea in the liver as his first project "because it appeared to be one of the simplest biosynthetic processes and one that occurs at a high rate." Using a liver tissue slice assay, Krebs and his research associate, Kurt Henseleit, showed that ornithine acted as a catalyst, as did citrulline and arginine, in the production of urea from ammonium and carbon dioxide. These results were published in a series of three papers that gave Krebs immediate recognition. His key insight was realizing that this pathway was a cycle and not a linear metabolic pathway. This mental preparation and his familiarity now with amino acid metabolism were key elements in his Nobel Prize-winning dissection of the TCA cycle in 1937. Remarkably, Krebs determined, with Hans Kornberg, a third metabolic cycle, the glyoxylate cycle, the glyoxylate bypass of the TCA cycle, in 1957.

in the TCA cycle. The other molecule that is recycled here is aspartate. The fumarate generated in the urea cycle in the cytosol is converted by cytosolic fumarase to malate, which is transported into the mitochondrial matrix by the malate-aspartate shuttle. Subsequently, malate is converted into oxaloacetate by malate dehydrogenase, generating the reducing equivalent NADH, which can generate ATP through oxidative phosphorylation. Thus, the ATP cost of the urea cycle is offset by the generation of this NADH molecule. The oxaloacaetate generated can be used as a substrate by aspartate aminotransferase to generate aspartate, which is transported back into the cytosol by the malate-aspartate shuttle to be used by the urea cycle. Hence, the urea and TCA cycles are metabolically linked through common metabolic intermediates.

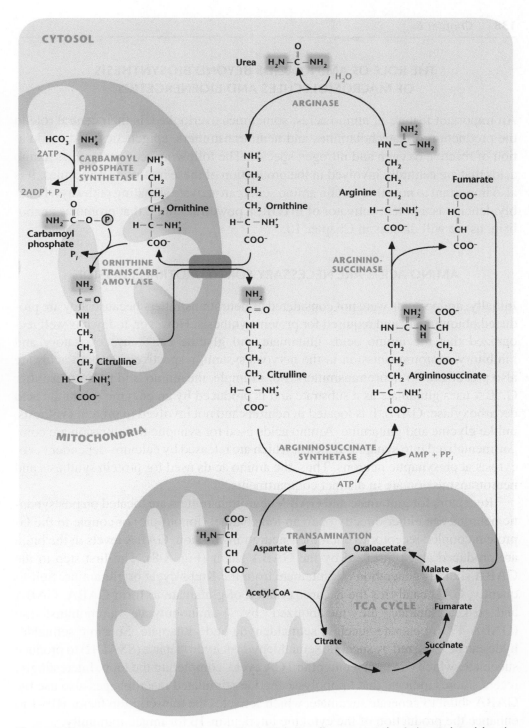

**Figure 8-7.** Overview of the urea cycle. The initial step for urea synthesis is catalyzed by the mitochondrial matrix enzyme carbamoyl phosphate synthetase I to incorporate the first nitrogen molecule. Subsequently, ornithine transcarbamoylase uses carbamoyl phosphate and ornithine as substrates to generate citrulline in the mitochondrial matrix. Next, citrulline is exported to the cytosol and converted into argininosuccinate by argininosuccinate synthetase. This reaction results in the incorporation of a second nitrogen atom from aspartate. Next, the enzyme argininosuccinase cleaves argininosuccinate to produce fumarate and arginine. Finally, the enzyme arginase converts arginine to urea and ornithine to complete the urea cycle.

## THE ROLE OF AMINO ACIDS BEYOND BIOSYNTHESIS OF MACROMOLECULES AND BIOENERGETICS

An important feature of amino acids, sometimes overlooked, is their central role in the production of catecholamines and neurotransmitters, epigenetics, and modulation of reactive oxygen and nitrogen species. The following sections discuss amino acids that are centrally involved in the production of these important molecules. It is also important to note that certain amino acids can activate signaling pathways. Notably, leucine is a potent activator of mTOR, a powerful kinase that promotes anabolism, as we will discuss in Chapter 10.

### AMINO ACIDS ARE NECESSARY FOR NEUROTRANSMISSION

Initially, amino acids were not considered as neurotransmitters because they are produced ubiquitously and required for protein synthesis. However, today it is well recognized that the amino acids glutamate and glycine can cause excitatory and inhibitory neurotransmission in the nervous system, respectively. Amino acids are also precursors for neurotransmitters. For example, the amino acid neurotransmitter GABA uses glutamate as a substrate and is produced by an enzyme (glutamic acid decarboxylase; GAD). It is located in neurons and not involved in protein synthesis, unlike glycine and glutamine. Amino acids used for synaptic transmission are compartmentalized into synaptic vesicles, which are released by calcium-dependent exocytosis at presynaptic neurons. Thus, the amino acids used for protein synthesis and neurotransmission are in distinct compartments.

Receptors for glutamate and GABA neurotransmitters are located on postsynaptic neurons can either directly open an ion channel (ionotropic) or couple to the G-protein coupled receptor to elicit their neurotransmission. GABA levels in the brain are produced and conserved by the GABA shunt (Fig. 8-8). The first step in the GABA shunt is generation of glutamate from the Krebs cycle or glutamine. Subsequently, GAD catalyzes the decarboxylation of glutamate to form GABA. GABA and $\alpha$-ketoglutarate are metabolized by 4-aminobutyrate aminotransferase (GABA-T) to generate succinic semialdehyde and glutamate. Succinic semialdehyde can be oxidized by succinic semialdehyde dehydrogenase (SSADH) to produce succinate, which can then reenter the TCA cycle, completing the loop. Interestingly, recent studies indicate that lipopolysaccharide-stimulated macrophages also use the GABA shunt to generate succinate, which activates the transcription factor HIF-1 to enhance the production of the cytokine interleukin-1$\beta$ for innate immunity.

The amino acid tryptophan is used as a precursor for generation of the neurotransmitter serotonin. Increased levels of serotonin are associated with feelings of happiness, whereas low-serotonin levels are linked to depression. The initial rate-limiting step in synthesis of serotonin is the conversion of L-tryptophan to 5-hydroxytryptophan by the enzyme L-tryptophan hydroxylase (Fig. 8-9). The

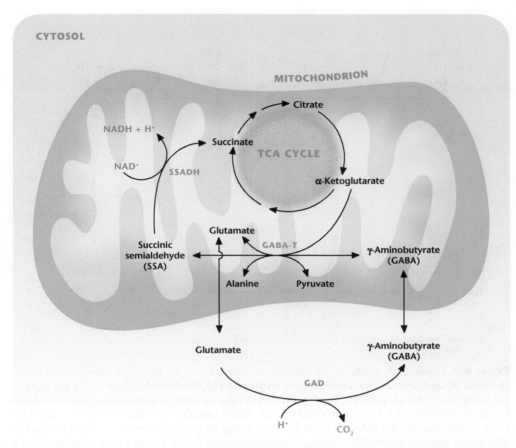

**Figure 8-8.** The GABA shunt produces and conserves GABA levels. The GABA shunt is initiated by GAD, catalyzing the decarboxylation of glutamate to form GABA. Subsequently, GABA and α-ketoglutarate are metabolized by GABA-T to generate succinic semialdehyde and glutamate. Succinic semialdehyde can be oxidized by SSADH succinate and can then reenter the Krebs cycle.

subsequent metabolic step is the decarboxylation of 5-hydroxytryptophan to generate 5-hydroxytryptamine (i.e., serotonin) by the enzyme aromatic amino acid decarboxylase. Interestingly, the generation of the sleep hormone melatonin is downstream from serotonin synthesis (Fig. 8-9). Serotonin *N*-acetyltransferase (arylalkylamine *N*-acetyltransferase) converts serotonin and acetyl-CoA into *N*-acetylserotonin, which becomes melatonin by acetylserotonin *O*-methyltransferase.

## TYROSINE METABOLISM GENERATES NEUROTRANSMITTERS, CATECHOLAMINES, AND MELANINS

Tyrosine is the precursor for the catecholamines norepinephrine, epinephrine, and dopamine (Fig. 8-10). Tyrosine hydroxylase converts tyrosine to generate dihydroxyphenylalanine (L-DOPA), a metabolic precursor to dopamine and dopaquinone.

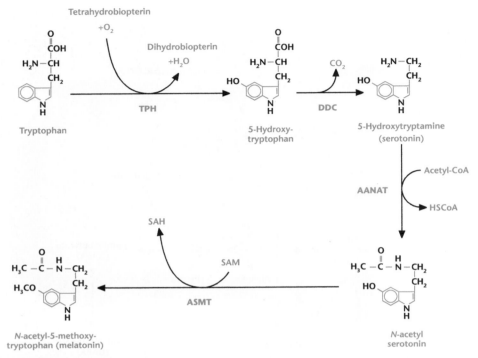

**Figure 8-9.** Tryptophan produces the neurotransmitter serotonin. The initial rate-limiting step in synthesis of serotonin is the conversion of L-tryptophan to 5-hydroxytryptophan by the enzyme L-tryptophan hydroxylase (TPH). Subsequently, aromatic L-amino acid decarboxylase (DDC) generates serotonin from 5-hydroxytryptophan. Serotonin can also generate melatonin. Arylalkylamine N-acetyltransferase (AANAT) adds an acetyl group to serotonin, resulting in the production of N-acetyl serotonin. Acetylserotonin O-methyltransferase (ASMT) adds a methyl group from S-adenosylmethionine (SAM) to N-acetyl serotonin, resulting in the formation of S-adenosylhomocysteine (SAH) and melatonin.

Tyrosine hydroxylase requires the enzyme cofactor tetrahydrobiopterin. L-DOPA is converted to dopamine by the enzyme aromatic amino acid decarboxylase. Dopamine is a potent neurotransmitter that is required for numerous brain functions. Notably, patients with Parkinson's disease, a debilitating condition marked by motor impairment and tremors, have few normal dopamine-producing cells in the midbrain area called the substantia nigra. Thus, L-DOPA is often prescribed to patients with Parkinson's disease to elevate their dopamine levels. There is also accumulating evidence that too-high levels of dopamine are observed in schizophrenia. The antipsychotic drugs for treatment work, in part, by limiting dopamine levels. Dopamine is also a precursor to epinephrine and norepinephrine, produced in the adrenal medulla. Catecholamines are associated with the fight-or-flight response and prepare the body to deal with environmental stress. The effects of catecholamines are associated with the sympathetic nervous system and increase blood pressure, heart rate, and blood glucose levels. Tyrosine is also a precursor to the production of the melanins,

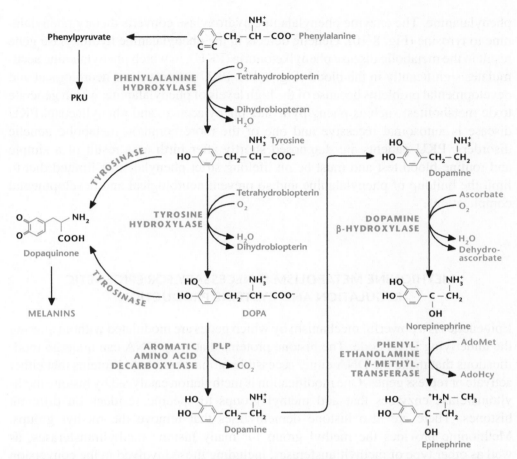

**Figure 8-10.** Tyrosine metabolism produces neurotransmitters, catecholamines, and melanins. Tyrosine hydroxylase converts tyrosine to generate dihydroxyphenylalanine (DOPA), which is converted to dopamine by the enzyme aromatic amino acid carboxylase. Dopamine generates norepinephrine through dopamine β-hydroxylase. Phenylethanolamine *N*-methyltransferase converts norepinephrine to epinephrine. Tyrosine is also a precursor for melanins through tyrosine or DOPA oxidation to dopaquinone by the enzyme tyrosinase. Phenylalanine hydroxylase produces tyrosine from phenylalanine. Genetic defects in the phenylalanine hydroxylase gene result in the metabolic disease phenylketonuria (PKU). PKU patients show high levels of phenylalanine that generate toxic metabolites, such as phenylpyruvate, resulting in neurological and developmental problems.

eumelanins, and pheomelanins by melanocytes to provide skin and hair pigmentation (Fig. 8-10). Eumelanins provide dark pigments, such as brown or black, and pheomelanins give rise to red or yellow pigmentation. The ratio of eumelanins and phomelanins in melanocytes dictates skin and hair pigmentation. People lacking the gene tyrosinase cannot generate pigments; thus, they exhibit alibinism.

Given the multiple roles of tyrosine metabolism, the maintenance of tyrosine levels within cells is vital. Tyrosine can be obtained from the diet or generated from phenylalanine. The enzyme phenylalanine hydroxylase converts dietary phenylalanine to tyrosine (Fig. 8-10). Genetic defects in the phenylalanine hydroxylase gene result in the metabolic disease phenylketonuria (PKU), in which phenylalanine accumulates significantly in the blood (Fig. 8-10). PKU patients show neurological and developmental problems because of the high levels of phenylalanine, which generate toxic metabolites, such as phenylpyruvate, phenylacetate, and phenyllactate. PKU disease is autosomal recessive and one of the more common metabolic genetic disorders. PKU patients are diagnosed shortly after birth as a result of a simple and routine blood test and must be on lifelong strict phenylalanine-limited diet to limit the buildup of phenylalanine and so prevent neurological and developmental complications.

## METHIONINE METABOLISM IS NECESSARY FOR EPIGENETIC REGULATION AND CYSTEINE PRODUCTION

Epigenetics is a powerful mechanism by which genes are modulated without altering the underlying DNA code. The histone proteins that wrap DNA can undergo modifications that make the DNA either accessible or inaccessible to proteins that either activate or repress genes. One modification is methylation catalyzed by histone methyltransferase enzymes that add methyl groups to specific residues on different histones. There are also histone demethylases that remove the methyl groups. Methionine provides the methyl group for many histone methyltransferases, as well as other type of methyltransferases, including those involved in the conversion of norepinephrine to epinephrine. These methyltransferases use $S$-adenosylmethionine (SAM). SAM is generated by condensation of ATP and methionine catalyzed by methionine adenosyltransferase. The methyl group ($CH_3$) is attached to the methionine sulfur atom in SAM. During the generation of SAM, all the ATP phosphates are lost so that only the adenosine component is attached to methionine (Fig. 8-11).

SAM is converted into $S$-adenosylhomocysteine (SAH) upon transferring its methyl group to DNA or proteins. SAH is then cleaved by adenosylhomocyteinase to yield homocysteine and adenosine. Methionine synthase can convert homocysteine back to methionine, a reaction that requires 5-MTHF as the methyl donor. The resulting THF can undergo a series of reactions, known as the folate cycle, to generate 5-MTHF (Fig. 8-11). Dietary folate provides THF. Betaine-homocysteine $S$-methyltransferase can also generate methionine by using homocysteine and betaine as substrates. Homocysteine can generate cysteine by a series of reactions known as *trans*-sulfuration (Fig. 8-11). Homocysteine condenses with serine to produce cystathionine, which is subsequently cleaved by cystathionine γ-lyase (also called cystathionase) to produce α-ketobutyrate and cysteine. α-Ketobutyrate is converted to propionyl-CoA and then, via a three-step process, to the TCA cycle

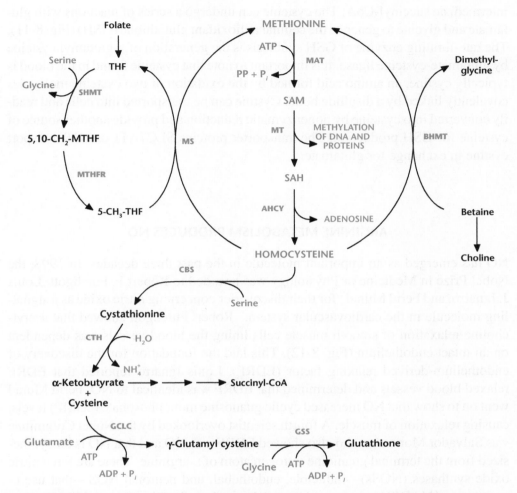

**Figure 8-11.** Methionine metabolism is used for methylation and cysteine production. Methionine adenosyltransferase (MAT) generates *S*-adenosylmethionine (SAM) using methionine and ATP as substrates. Methyltransferases (MTs) use SAM for methylation. The demethylation of SAM results in the formation of *S*-adenosylhomocysteine (SAH), which is converted into homocysteine and adenosine by adenosylhomocyteinase (AHCY). Methionine synthase can convert homocysteine back to methionine, a reaction that requires 5-methyl-tetrahydrofolate (5-MTHF) as the methyl donor. The resulting tetrahydrofolate (THF) is converted to 5,10-methylenetetrahydrofolate (5,10-CH$_2$-THF) by serine hydroxymethyl transferase (SHMT). 5,10-CH2-THF is then reduced to 5-MTHF by methylenetetrahydrofolate reductase (MTHFR). Betaine-homocysteine *S*-methyltransferase (BHMT) can also generate methionine from homocysteine by coupling betaine conversion to dimethylglycine. Homocysteine generates cysteine by a series of reactions known as *trans*-sulfuration. Serine condenses with homocysteine to generate cystathionine by cystathionine synthase (CBS). Subsequently, cystathionine is cleaved by cystathionine γ-lyase (CTH) to produce cysteine and α-ketobutyrate. Glutamate-cysteine ligase (GCLC) uses glutamate and cysteine as substrates to generate γ-glutamyl cysteine, which, in combination with glycine, is converted into glutathione by glutathione synthetase (GSS).

intermediate succinyl-CoA. The cysteine can undergo a series of reactions with glutamate and glycine to generate the cellular antioxidant glutathione (GSH) (Fig. 8-11). The rate-limiting enzyme of GSH synthesis is the generation of γ-glutamyl cysteine by glutamate-cysteine ligase. It is important to note that cysteine found in the blood is typically cystine, an amino acid formed by the oxidation of two cysteine molecules covalently linked by a disulfide bond. Cystine can be transported into cells and readily converted into cysteine by nonenzymatic reduction and provide another source of cysteine for GSH production. The transporter proteins SLC7A11 or xCT transport cystine in exchange for glutamate.

## ARGININE METABOLISM PRODUCES NO

NO has emerged as an important molecule in the past three decades. In 1998, the Nobel Prize in Medicine or Physiology was awarded to Robert F. Furchgott, Louis J. Ignarro, and Ferid Murad "for their discoveries concerning nitric oxide as a signalling molecule in the cardiovascular system." Robert Furchgott showed that acetylcholine relaxation of smooth muscle cells lining the blood vessels was dependent on an intact endothelium (Fig. 8-12). This laid the foundation for the discovery of endothelium-derived relaxing factor (EDRF). Louis Ignarro reported that EDRF relaxed blood vessels and determined that EDRF was identical to NO. Ferid Murad went on to show that NO increased cyclic guanosine monophosphate (cGMP) levels, causing relaxation of muscle. A fourth scientist overlooked by the Nobel Committee was Salvador Moncada, who also showed EDRF was NO and that NO is biosynthesized from the terminal guanidine nitrogen atom of L-arginine. There are three nitric oxide synthases (NOSs)—inducible, endothelial, and neuronal NOS—that use L-arginine and NADPH to generate NO and the product citrulline (Fig. 8-12). NO stimulation of soluble guanylate cyclase generates cGMP, which is normally broken down by phosphodiesterase type 5 (PDE5). The blockbuster drug Viagra used to treat erectile dysfunction (impotence) has the active ingredient sildenafil, which is a PDE5 inhibitor. Viagra prevents the breakdown of cGMP, which sustains blood vessel relaxation, improving blood flow to the penis and maintaining an erection. In addition to NO activation of soluble guanylate cyclase, NO can modulate cellular respiration by competing for the oxygen binding of cytochrome *c* oxidase in the mitochondrial electron transport chain. Furthermore, NO can cause protein posttranslational modification by incorporating the NO moiety into thiol groups to form *S*-nitrosothiol (SNO) (Fig. 8-12). The SNO modification results in a change in protein activity that is analogous to protein phosphorylation. This is a reversible reaction in which thioredoxin proteins can reverse protein SNO modification. This modification has been observed in all phylogenetic kingdoms and is thought to be an ancient signal transduction mechanism.

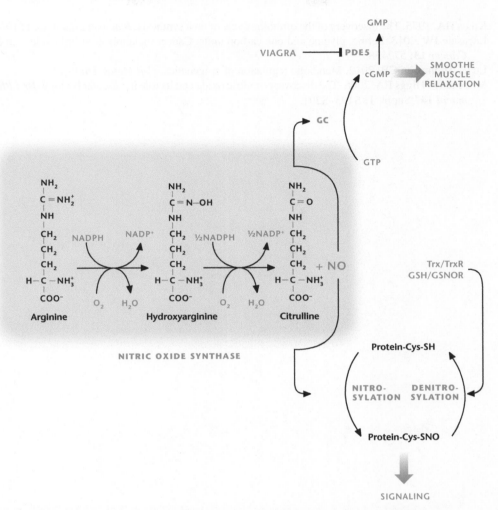

**Figure 8-12.** Arginine generates NO. NOs use L-arginine and NADPH to generate NO and the product citrulline. NO stimulation of soluble guanylate cyclase (GC) generates cGMP, which is normally broken down by phosphodiesterase type 5 (PDE5) to guanosine monophosphate (GMP). Viagra is a PDE5 inhibitor. NO can also cause protein posttranslational modifications by incorporating the NO moiety into thiol groups to form *S*-nitrosothiol (SNO). This modified thiol group can return to its reduced form through the actions of thioredoxin (TRX)/thioredoxin reductase (TrxR) or glutathione (GSH)/*S*-nitrosoglutathione reductase (GSNOR).

## ADDITIONAL READING

Altman LA. 1981. Sir Hans Krebs, Winner of Nobel for Research on Food Cycles, Dies. *The New York Times*, Obituary, December 9, 1981.

Hess DT, Stamler JS. 2012. Regulation by *S*-nitrosylation of protein post-translational modification. *J Biol Chem* **287:** 4411–4418.

Kornberg H. 2000. Krebs and his trinity of cycles. *Nat Rev Mol Cell Biol* **1:** 225–228.

Krebs HA. 1964. The citric acid cycle. Nobel lecture, December 11, 1953. In *Nobel Lectures, Physiology or Medicine 1942–1962*. The Nobel Foundation, Elsevier, Amsterdam.

Krebs HA. 1973. The discovery of the ornithine cycle of urea synthesis. *Biochem Education* **1:** 19–23.

Locasale JW. 2013. Serine, glycine and one-carbon units: Cancer metabolism in full circle. *Nat Rev Cancer* **13:** 572–583.

Lu C, Thompson CB. 2012. Metabolic regulation of epigenetics. *Cell Metab* **16:** 9–17.

Moncada S, Higgs EA. 2006. The discovery of nitric oxide and its role in vascular biology. *Br J Pharmacol* **147** Suppl. 1: S193–S201.

# 9

# Nucleotides

THE SINGLE MOST IMPORTANT EXPERIMENT OF THE 20th century was elucidation of the structure of DNA by James D. Watson and Francis H.C. Crick, along with Maurice Wilkins and Rosalind Franklin. Watson, Crick, and Wilkins were awarded the 1962 Nobel Prize in Physiology or Medicine "for their discoveries concerning the molecular structure of nucleic acids and its significance for information transfer in living material." (Franklin had died in 1958 and thus was not eligible for the award.) The discovery that there exists a heritable genetic code captured the imagination of the public and continues to have profound biological and social implications. In 1963, the surrealist artist Salvador Dali paid homage to the discovery in his painting, *Galacidalacidesoxyribonucleicacid.* Today, the discovery is celebrated and remembered in many ways, including my favorite, a plaque at The Eagle pub in Cambridge, United Kingdom, where, in 1953, Watson and Crick first publicly announced the discovery to their colleagues. Many decades later, DNA continues to be the center of biology. Furthermore, the discoveries of microRNA, noncoding RNA, and RNA interference have transformed biological science in the past two decades. Beyond genetics, nucleotides have important roles at the cellular level, including providing energy (ATP is a nucleotide!) and as signaling molecules (Fig. 9-1).

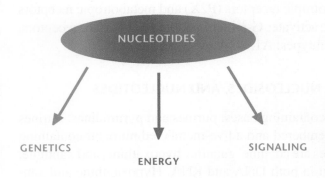

**Figure 9-1.** Overview of nucleotide metabolism. Nucleotides are key components of DNA and RNA (genetics), provide energy (ATP), and can activate signaling pathways.

## QUICK GUIDE TO NUCLEOTIDES

- Nucleosides have either a ribose or 2-deoxyribose bound to purine or pyrimidine. The addition of one or more phosphates to a nucleoside results in a nucleotide.

- Purines (adenine and guanine) are comprised of attached six-membered and five-membered nitrogen-containing rings.

- Pyrimidines (uracil, thymine, and cytosine) have only a six-membered nitrogen-containing ring.

- Ribonucleotide reductase (RR) generates deoxynucleoside diphosphate (dNDP) from ribonucleoside diphosphate (rNDPs). Nucleoside diphosphate (NDP) kinases use ATP to phosphorylate dNDP to produce deoxynucleoside triphosphates (dNTPs).

- Purine nucleotides synthesis begins with 5-phosphoribosyl-1-pyrophosphate (PRPP), which, through a series of reactions, generates the nucleotide inosine $5'$-monophosphate (IMP). Subsequently, IMP can be converted into either AMP or GMP through distinct reactions. AMP or GMP can be converted to ADP or GDP, respectively.

- Pyrimidine nucleotides synthesis begins with carbamoyl phosphate and aspartate generating the pyrimidine base orotate. Succeeding steps attach PRPP to orotate to generate orotate monophosphate (OMP), which is then decarboxylated to UMP. UMP generates UDP and UTP, which can generate CTP.

- Humans cannot break down the purine ring. The catabolism of purine nucleotides results in a uric acid. In contrast, the pyrimidine ring can be completely degraded. Catabolism of the pyrimidine nucleotides leads, ultimately, to β-alanine or β-aminoisobutyrate production, as well as $NH_3$ and $CO_2$.

- Nucletodies are signaling molecules that regulate multiple physiological processes, including neurotransmission and inflammation.

- ATP activates a family of ionotropic receptors (P2X) and metabotropic receptors (P2Y) Extracellular adenosine activates G-protein-coupled cell-surface receptors, which are divided into four subtypes: A1, A2A, A2B, and A3.

## NITROGEN BASES, NUCLEOSIDES, AND NUCLEOTIDES

There are two kinds of nitrogen-containing bases: purines and pyrimidines. Purines are comprised of attached six-membered and a five-membered nitrogen-containing rings (Fig. 9-2). The major purines are adenine, guanine, hypoxathine, and xanthine. Adenine and guanine are present in both DNA and RNA. Hypoxanthine and xanthine are not incorporated into the nucleic acids, but they are important intermediates in the synthesis and degradation of the purine nucleotides. Pyrimidines have only a

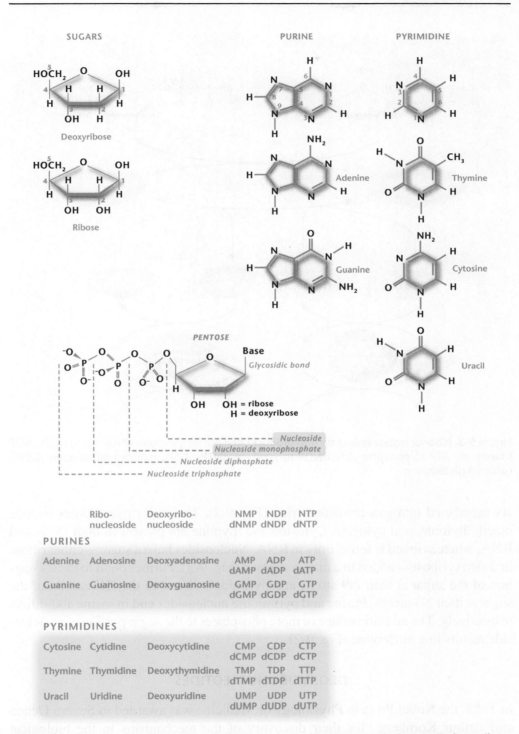

**Figure 9-2.** Structure of purines and pyrimidines. Nucleosides have either a ribose or 2-deoxyribose bound to purine or pyrimidine. The addition of one or more phosphates to a nucleoside results in a nucleotide. Purines (adenine and guanine) are compromised of attached six-membered and five-membered nitrogen-containing rings. Pyrimidines (uracil, thymine, and cytosine) have only a six-membered nitrogen-containing ring.

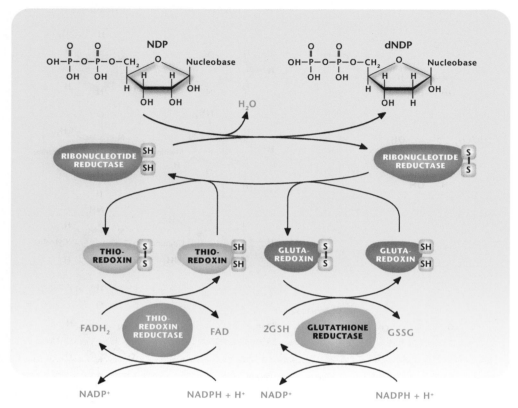

**Figure 9-3.** Ribonucleotide reductase (RR) produces dNDPs. RR generates dNDP from rNDPs. NDP kinases use ATP to phosphorylate dNDP to produce dNTPs. GSH, reduced glutathione; GSSG, oxidized glutathione.

six-membered nitrogen-containing ring (Fig. 9-2). The major pyrimidines include uracil, thymine, and cytosine. Cytosine and thymine are present in both DNA and RNA, whereas uracil is found only in RNA. Nucleosides have a sugar—either ribose or 2-deoxyribose—added to a nitrogen group (Fig. 9-3). Purines bond to the C1′ carbon of the sugar at their N9 atoms and pyrimidines bond to the C1′ carbon of the sugar at their N1 atoms. Purine and pyrimidine nucleosides end in -osine and -idine, respectively. The addition of one or more phosphates to the sugar portion of a nucleoside results in a nucleotide (Fig. 9-2).

## DEOXYRIBONUCLEOTIDES

In 1959, the Nobel Prize in Physiology or Medicine was awarded to Severo Ochoa and Arthur Kornberg "for their discovery of the mechanisms in the biological synthesis of ribonucleic acid and deoxyribonucleic acid." DNA and RNA polymerases use nucleotides to build DNA and RNA molecules. Most cells contain more RNA—messenger, ribosomal, and transfer RNAs—than DNA. Thus, nucleotide biosynthesis generates ample ribonucleotides. Because proliferating cells need to replicate their genomes, the production of deoxynucleotides is also necessary and begins with the reduction of rNDPs to generate dNDP by RR (Fig 9-3). RR has a large R1 subunit that is a dimer of monomeric molecular mass 90 kDa, whereas the smaller R2 subunit is a dimer of monomeric molecular mass 45 kDa. This multifunctional enzyme contains redox-active thiol groups that participate in the transfer of electrons during the reduction reactions. RR becomes oxidized as rNDP and is reduced to dNDP. RR is reduced to its original state by either thioredoxin or glutaredoxin, which in turn become oxidized and are regenerated to the reduced form by the enzymes thioredoxin reductase and glutathione reductase, respectively. The ultimate source of the electrons driving these reactions is NADPH (see also Chapter 5).

RR is the only enzyme used in the generation of all the deoxyribonucleotides, and so its activity is tightly regulated to ensure stable production of all four of the dNTPs required for DNA replication. Each R1 monomer contains two distinct binding sites for deoxynucleoside triphosphates, which act as allosteric regulators of the enzymatic activity. The binding of dATP to allosteric sites, known as the activity sites, on the enzyme inhibits the overall catalytic activity of the enzyme. This prevents the reduction of any of the four NDPs to effectively block DNA synthesis. In contrast, ATP bound to these sites activates the enzyme. Thus, the ATP/dATP ratio modulates the activity of the enzyme to ensure that sufficient dNTPs are produced for DNA synthesis when needed. In the absence of this regulatory mode, ribonucleotides needed for RNA synthesis would be depleted by needlessly being converted into deoxynucleotides.

Another allosteric site, known as a substrate-specific site, is responsible for maintaining balanced nucleotide pools, for optimal fidelity of DNA replication. Deoxythymidine triphosphate (dTTP) binding at the specificity sites causes a conformational change that allows reduction of GDP to dGDP at the catalytic site. The binding of ATP or dATP induces reduction of cytidine 5′-diphosphate (CDP) and uridine 5′-diphosphate (UDP) to dCDP and dUDP, respectively. The binding of 2′-deoxyguanosine 5′-triphosphate (dGTP) causes reduction of adenosine 5′-diphosphate (ADP) to dADP. Once a dNDP is generated by RR, that particular dNDP can undergo phosphorylation to produce dNTPs, a reaction catalyzed by the NDP kinases using ATP as the phosphate donor. There are also nucleoside monophosphate (NMP) kinases that catalyze ATP-dependent reactions of the type (d)NMP + ATP ↔ (d)NDP + ADP. The activity of the NDP kinases is higher than that of the NMP kinases to maintain a high intracellular level of (d)NTPs relative to that of (d)NDPs.

## PURINE NUCLEOTIDE BIOSYNTHESIS

Purine nucleotides synthesis begins with PRPP, which, through a series of reactions, generates the nucleotide IMP. The first reaction is catalyzed by ribose 5-phosphate pyrophosphokinase to generate PRPP from ribose 5-phosphate and the two high-energy phosphates of ATP. Ribose 5-phosphate serves as a platform for the construction of the purine ring structure. PRPP is not committed just to purine synthesis; it is also used for other processes, including pyrimidine and histidine biosynthesis. The second step, catalyzed by PRPP glutamylamidotransferase, commits PRPP to purine synthesis and is tightly regulated. In this step, there is a loss of the PRPP pyrophosphate and the addition of an amino group from glutamine. Subsequently, there are nine additional enzymatic reactions that result in the formation of IMP, which contains the nitrogen base hypoxanthine. The pathway first builds the five-membered ring and then the six-membered ring (Fig. 9-4). The amino acids glutamine, aspartate, and glycine serve as nitrogen donors, as ammonium ($NH_4^+$) is not a nitrogen donor in biosynthetic reactions. The synthesis of IMP requires five moles of ATP, two moles of glutamine and formate, and one mole of glycine, $CO_2$, and aspartate. The formyl moieties are transported on tetrahydrofolate (THF) in the form of $N^{10}$-formyl-THF. The rate-limiting steps in purine biosynthesis are the first two steps of the pathway (Fig. 9-5). The first step catalyzed by ribose-5-pyrophosphokinase is a feedback-inhibited reaction by ADP and GDP. The second reaction is catalyzed by PRPP amidotransferase. This reaction is also feedback inhibited allosterically by binding of AMP, GMP, and IMP. Conversely, the activity of PRPP amidotransferase is stimulated by PRPP.

IMP can be converted into either AMP or GMP through distinct reactions (Fig. 9-6). The formation of AMP requires GTP and production of GMP requires ATP. This allows for control of AMP or GMP depending on the availability of GTP or ATP, respectively. The accumulation of GTP results in AMP synthesis from IMP at the cost of GMP production. Conversely, excess ATP favors GMP production at the expense of AMP synthesis. Furthermore, AMP or GMP accumulation inhibits adenylosuccinate synthetase or IMP dehydrogenase to prevent further production of AMP or GMP, respectively (Fig. 9-6). Subsequently, AMP or GMP can be converted to ADP or GDP by adenylate kinase or guanylate kinase using ATP, respectively (Fig. 9-5). Because free ATP is present in higher levels than any other nucleotide, it is usually used as the phosphate donor in these reactions.

## PURINE NUCLEOTIDE DEGRADATION

Humans cannot break down the purine ring. The catabolism of purine nucleotides results in uric acid, which has low solubility in water and is excreted as sodium urate crystals in the urine (Fig. 9-7). Purine catabolism does not alter the basic purine structure. The primary pathway is initiated by conversion of AMP to IMP or adenosine to generate inosine, which subsequently makes hypoxanthine. GMP can be converted to guanosine and, subsequently, into guanine. Both guanine and hypoxanthine generate xanthine. The final step involves converting xanthine to uric acid by xanthine oxidase, a step that produces hydrogen peroxide.

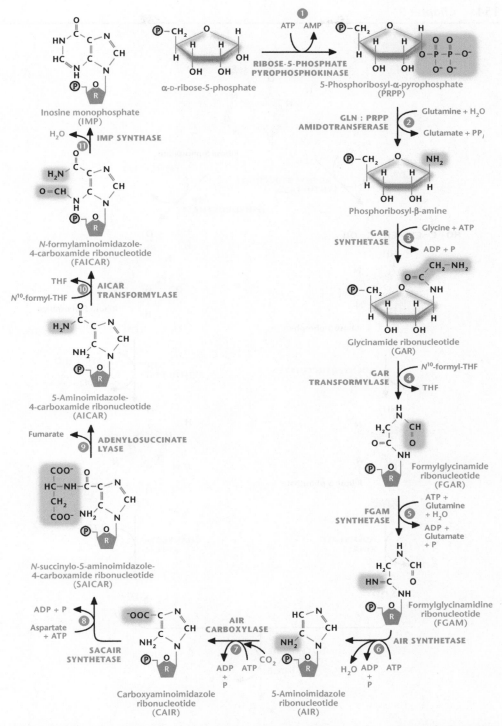

**Figure 9-4.** The purine synthesis pathway. The purine ring is built on the ribose sugar 5-phosphoribosyl-1-pyrophosphate (PRPP) as the base. Purine nucleotides synthesis begins with PRPP, which, through a series of reactions, generates the nucleotide, inosine 5'-monophosphate (IMP). (Adapted from Garrett and Grisham 1999.)

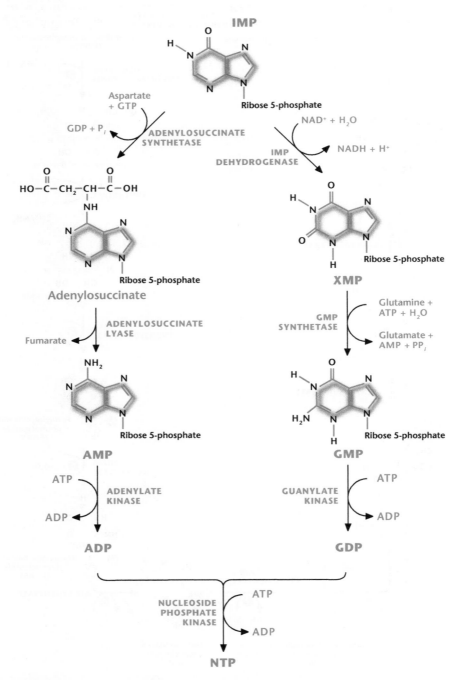

**Figure 9-5.** Purine nucleotide production. IMP can be converted into either AMP or GMP through distinct reactions. AMP or GMP can be converted to ADP or GDP by adenylate kinase or guanylate kinase, respectively.

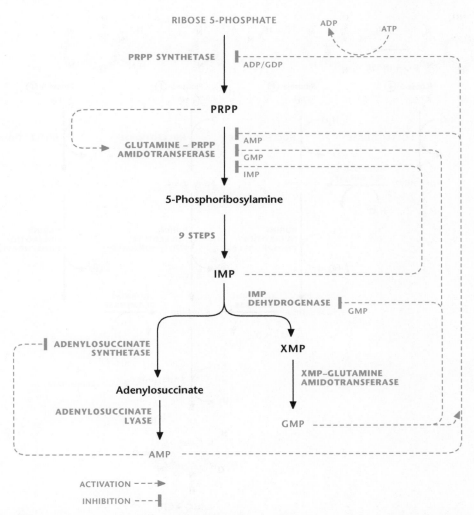

**Figure 9-6.** Regulation of purine biosynthesis. The first two reactions of the purine biosynthesis pathway are rate-limiting steps, which are regulated. (Adapted, with permission, from Nelson and Cox 2013, © W.H. Freeman and Company.)

## PURINE NUCLEOTIDE SALVAGE

Nucleotides can also be synthesized from the purine bases and purine nucleosides in a series of steps referred to as salvage pathways. Phosphoribosylation converts the free purine bases adenine, guanine, and hypoxanthine to their corresponding nucleotides. Two key transferase enzymes are involved in the salvage of purines are adenosine phosphoribosyltransferase (APRT), which converts adenine to AMP, and hypoxanthine-guanine phosphoribosyltransferase (HGPRT), which converts hypoxanthine to IMP or guanine to GMP by using PRPP as a substrate (Fig. 9-8). By using PRPP as a substrate, the salvage pathway decreases the levels of PRPP, and therefore

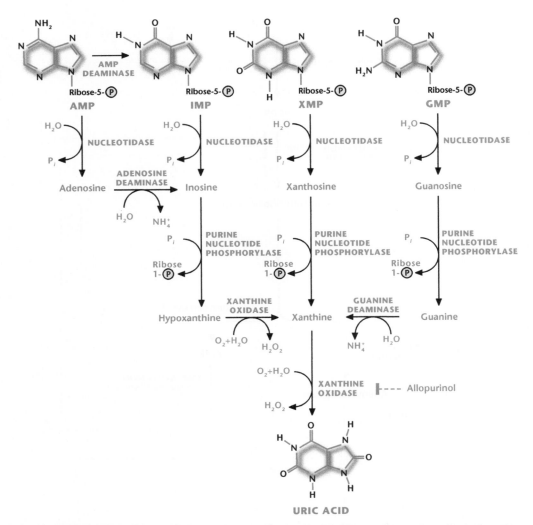

**Figure 9-7.** Catabolism of purines. Humans cannot break down the purine ring of purine nucleotides. The catabolism of purine nucleotides results in a uric acid, which has low solubility in water and is excreted as sodium urate crystals in the urine. Allopurinol inhibits xanthine oxidase and is used to treat gout.

diminishes the rate of de novo purine synthesis. The de novo purine synthesis and salvage pathways are tightly coupled based on the availability of PRPP.

One important salvage pathway is the purine nucleotide cycle, which produces the TCA-cycle intermediate fumarate as a mechanism to boost the TCA cycle's ability to generate more NADH for the production of ATP (Fig. 9-9). An increase in muscle activity creates a high-metabolic demand for ATP. An increase in AMP levels is a signal that energy generation by the TCA cycle and electron transport chain is insufficient to keep up with metabolic demand. This cycle is very important in muscle cells because they lack most of the enzymes of the major anapleurotic

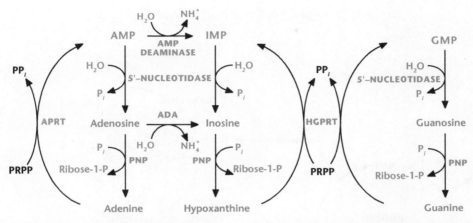

**Figure 9-8.** Purine salvage pathway. Two key transferase enzymes involved in the salvage of purines are adenosine phosphoribosyltransferase (APRT), which converts adenine to AMP, and hypoxanthine-guanine phosphoribosyltransferase (HGPRT), which converts hypoxanthine to IMP or guanine to GMP by using 5-phosphoribosyl-1-pyrophosphate (PRPP) as a substrate.

reactions (i.e., those that replenish pathway intermediates). Thus, muscle replenishes TCA-cycle intermediates through fumarate generated by the purine nucleotide cycle. Myoadenylate deaminase is the muscle-specific isoenzyme of AMP deaminase, and deficiencies in myoadenylate deaminase lead to postexercise fatigue and cramping.

---

**BOX 9-1. GOUT**

Clinical problems associated with nucleotide metabolism in humans are predominantly the result of abnormal catabolism of the purines and not pyrimidine. Remember, pyrimidine ring structure can be fully degraded, whereas the purine ring degradation results in the relatively insoluble compound uric acid. Elevated levels of uric acid can result in the uric acid crystal formation in the joints that characterizes gout or can form kidney stones. Gout is initiated by high levels of uric acid in the blood (hyperuricemia), as a result of either the overproduction or underexcretion of uric acid. The uric acid crystals engage the caspase-1-activating inflammasome, resulting in the production of interleukin-1β and IL-18. This produces severe inflammation and arthritis in the joints. In the majority of people, hyperuricemia is caused by underexcretion of uric acid, whereas a less common cause of hyperuricemia is from the overproduction of uric acid. Several identified mutations in the gene for PRPP synthetase result in increased availability of PRPP, causing overproduction of purines. Rare inherited disorders, such as Lesch–Nyhan syndrome, caused by a deficiency in HGPRT also cause hyperuricemia by decreasing salvage of hypoxanthine and guanine, therefore, increasing availability of PRPP. Administering allopurinol, a structural analog of hypoxanthine that inhibits xanthine oxidase, can treat gout. Allopurinol increases the levels of the more soluble metabolites xanthine, hypoxanthine, and guanine compared with uric acid.

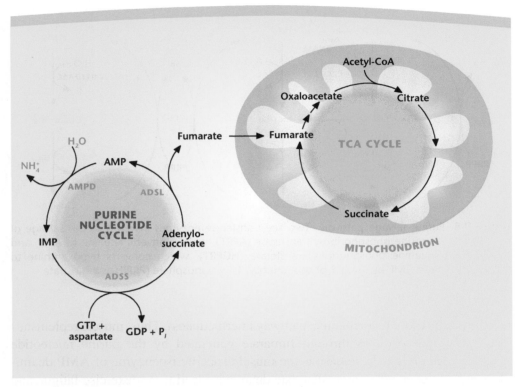

**Figure 9-9.** Purine nucleotide cycle. Purine nucleotide pathway converts AMP into IMP via AMP deaminase (AMPD). Next, adenylosuccinate synthetase (ADSS) converts IMP into adenylosuccinate using aspartate and GTP as substrates. Subsequently, adenylosuccinate lyase (ADSL) converts IMP into fumarate and AMP. Fumarate enters the TCA cycle.

## PYRIMIDINE NUCLEOTIDE BIOSYNTHESIS

Pyrimidine synthesis differs from purine synthesis in that the pyrimidine ring is fully synthesized before being attached to the ribose sugar, whereas the purine ring is built on the ribose sugar (PRPP) as the base. Overall synthesis of pyrimidines is a less-complex process than purines primarily because of the simpler six-membered ring base structure (Fig. 9-10). Carbamoyl phosphate and aspartate contribute to the ring structure. The carbamoyl phosphate is derived from 1 mole of glutamine, 2 moles of ATP, and 1 mole of bicarbonate within the cytosol. Note that the urea cycle also uses carbamoyl phosphate, but it is derived from ammonia and bicarbonate in the mitochondrion. The urea cycle reaction is catalyzed by carbamoyl phosphate synthetase I, whereas the pyrimidine nucleotide precursor is generated by carbamoyl phosphate synthetase II (CPS-II). Subsequently, carbamoyl phosphate funnels into the pyrimidine nucleotide biosynthesis pathway through the action of aspartate transcarbamoylase, which is the rate-limiting step in pyrimidine biosynthesis. The next steps build the first fully formed pyrimidine base orotate. PRPP attaches to oroate to

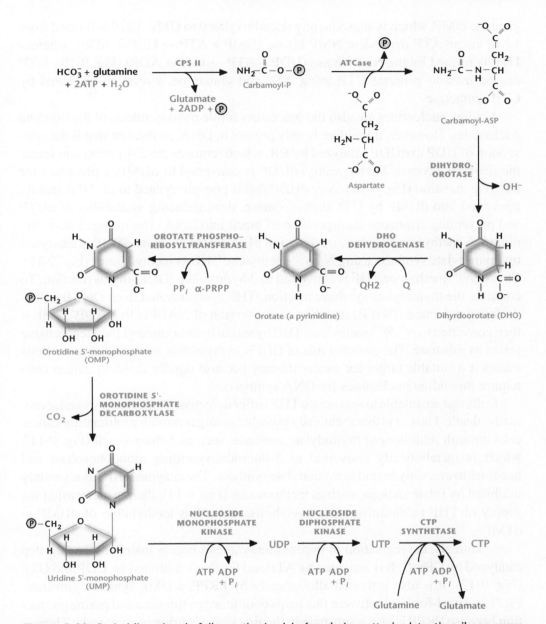

**Figure 9-10.** Pyrimidine ring is fully synthesized before being attached to the ribose sugar. Pyrimidine nucleotides synthesis begins with carbamoyl phosphate and aspartate generating the pyrimidine base orotate. Next, steps attach PRPP to oroate to generate orotate monophosphate (OMP), which is subsequently decarboxylated to UMP. UMP generates UDP and UTP, which can generate CTP. (Adapted from Garrett and Grisham 1999.)

generate OMP, which is subsequently decarboxylated to UMP. UDP is formed from UMP via an ATP-dependent NMP kinase (UMP + ATP → UDP + ADP), whereas UTP is formed by the NDP kinase (UDP + ATP → UTP + ADP) (Fig. 9-10). UTP can be used to generate CTP using ATP and glutamine, a reaction catalyzed by CTP synthetase.

Uridine nucleotides are also the precursors for de novo synthesis of the thymine nucleotides. However, thymidine is only present in DNA, so the first step is the conversion of UDP to dUDP catalyzed by RR, which removes the 2′-hydroxyl, to create the deoxynucleotide. Subsequently, dUDP is converted to dUMP, a precursor for dTMP generation (Fig. 9-11). Any dUDP that is phosphorylated to dUTP is rapidly converted into dUMP by UTP diphosphatase, thus, reducing availability of dUTP and preventing erroneous incorporation of uracil into DNA. The nitrogen base thymine is 5-methyl uracil. Thus, methylation of dUMP to generate dTMP is catalyzed by thymidylate synthase using $N^5,N^{10}$-methylene THF as methyl donor (Fig. 9-11). The $N^5,N^{10}$-methylene THF is converted to dihydrofolate (DHF) in this reaction. To continue the thymidylate synthase reaction, THF is regenerated from DHF by dihydrofolate reductase (DHFR) coupling to conversion of NADPH to NADP$^+$. THF is then converted to $N^5,N^{10}$-methylene THF by serine hydroxymethyl transferase using serine as substrate. The essential role of DHFR in thymidine nucleotide biosynthesis makes it a suitable target for cancer therapy because rapidly dividing cancer cells require thymidine nucleotides for DNA synthesis.

Cells that are unable to regenerate THF suffer defective DNA synthesis and, eventually, death. Thus, it is therapeutically possible to target rapidly proliferating cancer cells through inhibitors of thymidylate synthase, such as 5-fluorouracil (Fig. 9-11), which is metabolically converted to 5-fluorodeoxyuridine monophosphate and becomes irreversibly bound to thymidylate synthase. The enzyme DHFR is reversibly inhibited by folate analogs, such as methotrexate (Fig. 9-11), thereby decreasing the supply of THF to diminish purine synthesis, as well as methylation of dUMP to dTMP.

In animals, the regulation of pyrimidine synthesis occurs mainly at the first step catalyzed by CPS-II. It is activated by ATP and feedback inhibited by UDP and UTP (Fig. 9-12). It is also activated allosterically by PRPP, a UMP synthase substrate. PRPP is a feed-forward activator that helps coordinate pyrimidine and purine production because PRPP is also a substrate in the first step of purine biosynthesis. There is also regulation of orotidine 5′-phosphate decarboxylase, which is competitively inhibited by UMP and, to a lesser degree, CMP. Finally, CTP synthase is feedback-inhibited by CTP.

## CATABOLISM OF PYRIMIDINE NUCLEOTIDES

Unlike the purine ring, the pyrimidine ring can be completely degraded. Pyrimidine bases are water soluble, and the pyrimidine ring can undergo degradation to

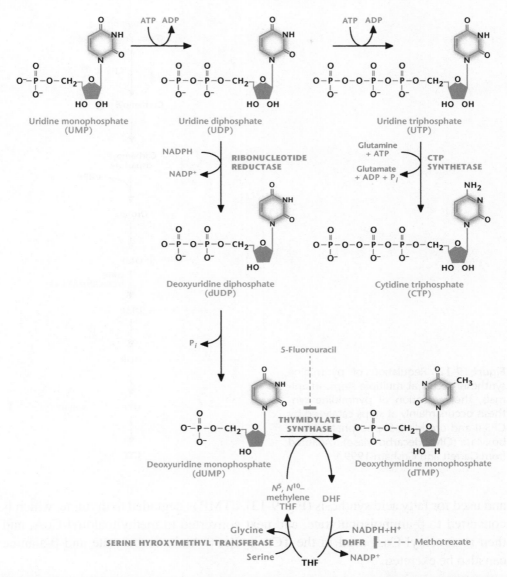

**Figure 9-11.** Uridine nucleotides are also the precursors for de novo synthesis of the thymine nucleotides. UDP to dUDP is catalyzed by ribonucleotide reductase. Subsequently, dUDP is converted to dUMP. The methylation of dUMP to generate dTMP is catalyzed by thymidylate synthase using $N^5,N^{10}$-methylene THF as methyl donor.

metabolites that can feed into metabolic pathways, such as the TCA cycle. Thus, pyrimidine salvage pathways are not as significant as purine salvage pathways. Catabolism of the pyrimidine nucleotides leads ultimately to β-alanine or β-aminoisobutyrate production, as well as $NH_3$ and $CO_2$. UMP and CMP are degraded to uracil and cytosine, which generate β-alanine, which is subsequently converted to malonyl-CoA

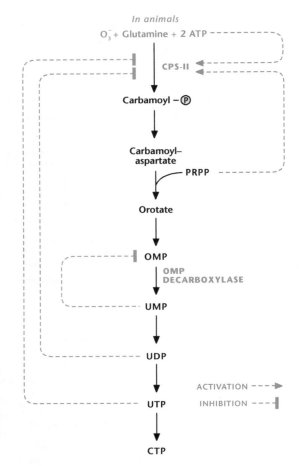

*In animals*

$O_3^-$ + Glutamine + 2 ATP

CPS-II

Carbamoyl – $\text{\textcircled{P}}$

Carbamoyl–
aspartate

PRPP

Orotate

OMP

OMP
DECARBOXYLASE

UMP

UDP

UTP

CTP

ACTIVATION

INHIBITION

**Figure 9-12.** Regulation of pyrimidine synthesis occurs at multiple steps. In animals, the regulation of pyrimidine synthesis occurs mainly at steps catalyzed by CPS-II and orotidine 5′-phosphate decarboxylase (OMP decarboxylase). (Adapted from Garrett and Grisham 1999.)

and used for fatty acid synthesis (Fig. 9-13). dTMP is degraded to thymine, which is converted to β-aminoisobutyrate, and next converted to methylmalonyl-CoA, and then to succinyl-CoA used in the TCA cycle. β-Aminoisobutyrate and β-alanine can also be excreted.

## NUCLEOTIDES AND SIGNALING

The cyclic nucleotides cAMP and cGMP are well-recognized signaling molecules. However, nucleotides themselves, as well as their breakdown products, have also emerged as important signaling molecules interacting with specific receptors (Fig. 9-14). Experiments dating back more than 50 years suggested that some nerves release ATP, and this later led to the proposal of "purinergic nerves." Today, it is well recognized that extracellular ATP can regulate multiple physiological processes, including neurotransmission and inflammation, by activating nucleotide receptors called P2 purinoreceptors. ATP activates a family of P2X and P2Y receptors,

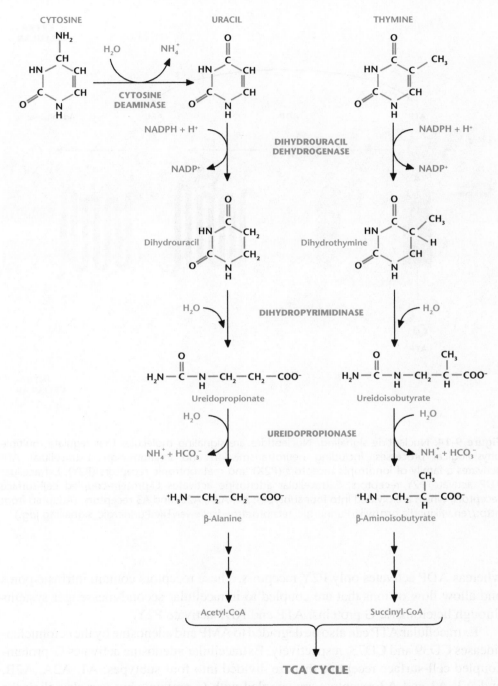

**Figure 9-13.** Catabolism of pyrimidine rings. Pyrimidine rings of cytosine and thymine can be completely degraded. Catabolism of the cytosine and thymine leads ultimately to β-alanine or β-aminoisobutyrate production, as well as $NH_3$ and $CO_2$. (Adapted from Moran et al. 2012, p. 570, by permission of Pearson Education Inc., Upper Saddle River, New Jersey.)

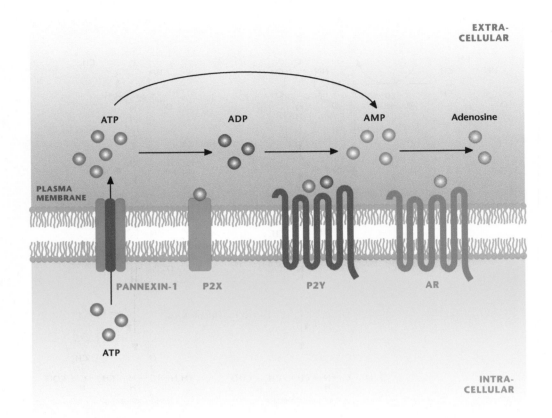

**Figure 9-14.** Nucleotide signaling. Nucleotides are signaling molecules that regulate multiple physiological processes, including neurotransmission and inflammation. Extracellular ATP activates a family of ionotropic receptors (P2X) and metabotropic receptors (P2Y). Extracelluar ADP activate P2Y receptors. Extracellular adenosine activates G-protein-coupled cell-surface receptors, which are divided into four subtypes: A1, A2A, A2B, and A3 receptors. (Adapted from http://en.wikipedia.org/wiki/Purinergic_receptor#mediaviewer/File:Purinergic_signalling.jpg.)

whereas ADP activates only P2Y receptors. These receptors contain intrinsic pores and allow flow of ions that are coupled to intracellular second-messenger systems through heteromeric G proteins. ATP and ADP activate P2Y.

Extracellular ATP can also be degraded to AMP and adenosine by the ectonucleotideases CD39 and CD73, respectively. Extracellular adenosine activates G-protein-coupled cell-surface receptors that are divided into four subtypes: A1, A2A, A2B, and A3. A1 and A3 receptors are coupled with G proteins, but can also elicit the release of calcium ions from intracellular stores. A2A and A2B receptors stimulate adenylyl cyclase by Gs proteins. A2B receptors can also activate phospholipase C through Gq proteins. All the adenosine receptors can activate mitogen-activated protein kinase (MAPK) pathways, including extracellular signal-regulated kinase

## BOX 9-2. CAFFEINE

Caffeine is a purine that was first synthesized from raw materials by the German chemist Hermann Emil Fischer in 1895. Fischer was well known for discovering purines and awarded the Nobel Prize in Chemistry in 1902 "in recognition of the extraordinary services he has rendered by his work on sugar and purine syntheses." Today, caffeine is unquestionably the most widely consumed substance that influences daily behavior. There is general perception that caffeine is "bad." However, like alcohol, there are data to indicate that moderate consumption of caffeinated drinks can be beneficial. Recent case-control and prospective studies have linked caffeine consumption with a reduced risk of type 2 diabetes, chronic liver cirrhosis, and neurodegenerative diseases, such as Parkinson and Alzheimer diseases, dementia, and type 2 diabetes. Furthermore, long-term caffeine use is not associated with increased morbidity. The main pharmacological target of caffeine is to antagonize adenosine receptors, particularly $A_{2A}$ receptors. These observations, along with laboratory-based experiments linking adenosine to inflammation, cancer, and neurodegeneration, have promoted the development of specific adenosine receptor antagonists, such as Istradefylline, that have undergone clinical trials for Parkinson disease.

1, -2, Jun amino-terminal kinase, and p38 MAPK. Ultimately, adenosine receptor signaling depends on the level of extracellular adenosine levels. Adenosine deaminase degrades extracellular adenosine, whereas cell-membrane-embedded nucleoside transporters shuttle extracellular adenosine into the intracellular space, thereby decreasing adenosine receptor activity.

## REFERENCES

Garrett RH, Grisham CM. 1999. *Biochemistry*, 2nd ed. Thomson Brooks/Cole, Belmont, CA.

Moran LA, Horton RA, Scrimgeour G, Perry M. 2012. *Principles of biochemistry*, 5th ed. Pearson, Glenville, IL.

Nelson DL, Cox MM. 2013. *Lehninger principles of biochemistry*, 6th ed. WH Freeman, New York.

## ADDITIONAL READING

Antonioli L, Blandizzi C, Pacher P, Haskó G. 2013. Immunity, inflammation and cancer: A leading role for adenosine. *Nat Rev Cancer* **13:** 842–857.

Freedman ND, Park Y, Abnet CC, Hollenbeck AR, Sinha R. 2012. Association of coffee drinking with total and cause-specific mortality. *N Engl J Med* **366:** 1891–1904.

Idzko M, Ferrari D, Eltzschig HK. 2014. Nucleotide signalling during inflammation. *Nature* **509:** 310–317.

Neogi T. 2011. Clinical practice. Gout. *N Engl J Med* **364:** 443–452.

Nordlund P, Reichard P. 2006. Ribonucleotide reductases. *Annu Rev Biochem* **75:** 81–70.

# 10

# Signaling and Metabolism

THE PAST NINE CHAPTERS COVERED THE basic aspects of various anabolic and catabolic pathways. In mammalian cells, these pathways are tightly regulated by availability of nutrients and stimulation of signaling pathways by cell-surface receptors engaged by extracellular factors. Mammalian cells take up nutrients under the instruction of these signaling pathways, unlike yeast cells, which take up nutrients in a cell-autonomous manner. These signaling pathways maintain homeostasis by allowing sufficient nutrient uptake to maintain metabolic demand. In the presence of ample nutrients and certain growth factors, cells grow in size and at times this is coupled with cell proliferation. If a particular nutrient (amino acid, oxygen, or glucose) becomes depleted, then cells have nutrient-sensing mechanisms that switch from an anabolic program to a catabolic program. Similarly, if nutrients are abundant, but the extracellular factors that activate signaling pathways for nutrient uptake are absent, then cells also undergo a catabolic program. By decreasing anabolism, the cells diminish their metabolic demand, and by activating catabolic pathways, they try to increase metabolic supply with the ultimate goal of maintaining an energy charge ratio that sustains survival (see Chapter 2). Autophagy ("self-eating") is a catabolic metabolic program that is activated when intracellular nutrient uptake or availability is diminished. Autophagy generates intracellular amino acids and other substrates that can be used to generate ATP. Cells cannot forever sustain autophagy because eventually there is nothing left to catabolize and the cells die of bioenergetic collapse. Amazingly, before this collapse, the reintroduction of intracellular nutrients allows the cells to grow back to their original size.

This chapter focuses on signaling pathways that promote nutrient uptake, respond to changes in nutrients, and regulate autophagy. Two critical metabolic pathways are glycolysis and mitochondrial oxidative metabolism (see Chapters 3 and 4); thus, we will discuss signaling pathways that are activated when glucose or oxygen levels diminish. Finally, the chapter focuses on how metabolites can confer posttranslational modifications (PTMs), such as acetylation, to regulate enzymes within metabolic pathways.

## SIGNAL TRANSDUCTION REGULATES UPTAKE AND USE OF NUTRIENTS

Extracellular factors engage their receptors to activate signal transduction pathways that concomitantly activate anabolic metabolic pathways and transporters for nutrient uptake in mammalian cells (Fig. 10-1). Insulin is perhaps the most potent anabolic factor. It stimulates the insulin receptor, a member of the cell membrane receptor

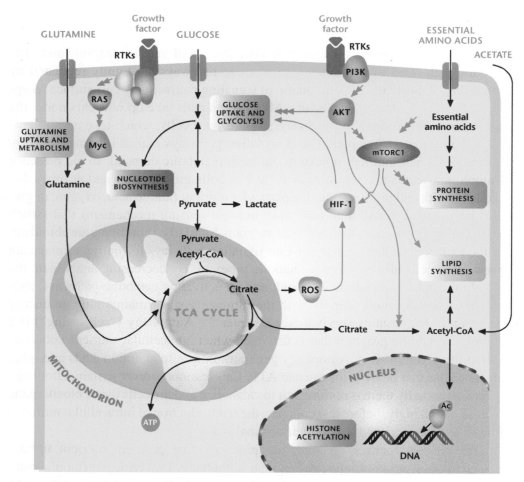

**Figure 10-1.** Signal transduction pathways regulate metabolism. In multicellular organisms, cells use growth factors to uptake nutrients. Growth factors through receptor tyrosine kinases (RTKs) activate the PI3K to stimulate AKT and mTORC1, which promotes glycolysis and lipid synthesis. mTORC1 signaling responds to levels of essential amino acids to stimulate protein synthesis. The HIF-1 transcription factor further promotes glycolysis. RTK activation also activates Myc transcription factor, which enhances both glycolysis and mitochondrial metabolism. Myc stimulates glutamine metabolism necessary to sustain mitochondrial metabolism and nucleotide biosynthesis. Mitochondrial ROS and acetyl-CoA production promote HIF-1 activation and histone acetylation, respectively, as examples of metabolism feeding back on signaling. (Adapted, with permission, from Ward and Thompson 2012.)

tyrosine kinases (RTKs), resulting in uptake of glucose, amino acids, and fatty acids into cells and promotes synthesis of lipids, proteins, and carbohydrates. The RTKs stimulate a diverse series of signaling pathways, including the PI3K/AKT/mTOR and RAS/RAF/MEK pathways. RTK activates the lipid kinase phosphoinositide 3-kinase (PI3K), which phosphorylates membrane lipid phosphatidylinositol 4,5-bisphosphate (PIP2) to generate phosphatidylinositol 3,4,5-trisphosphate (PIP3). PIP3 recruits serine/threonine-specific protein kinase AKT, also known as protein kinase B, to the cell membrane, where it is activated by phosphorylation by phosphoinositide-dependent kinase 1 (at threonine 308) and the mechanistic target of rapamycin complex 2 (mTORC2 at serine 473). The major output of AKT signaling is the activation of mechanistic target of rapamycin complex 1 (mTORC1), which increases multiple anabolic pathways, including protein and lipid synthesis (discussed in the next section). AKT, through mTORC1-independent and -dependent pathways, can increase both the activity and expression of glucose and amino acid transporters at the surface. Furthermore, AKT stimulates the activity of multiple glycolytic enzymes, including hexokinase, the first rate-limiting step of glycolysis. AKT, through mTORC1, can also increase the translation of HIF1α protein. HIF1α can heterodimerize with HIF1β to generate the transcriptionally active HIF1, which increases glucose transporters and glycolytic enzymes.

Additional transcriptional induction of metabolic enzymes comes from stabilization of the MYC protein by the RAS (rat sarcoma viral oncogene) activation of ERK mitogen-activated protein kinase. The newly synthesized MYC protein undergoes ubiquitination and degradation, but can be stabilized when phosphorylated at serine 62 (Ser62) by ERK activity. Stabilized MYC protein heterodimerizes with the MAX protein to activate transcription of enzymes involved in glycolysis and mitochondrial metabolism, as well as lipid, nucleotide, amino acid, and carbohydrate metabolism. In summary, signal transduction controls metabolism through posttranslational regulation by AKT and mTORC1 coupled with transcriptional regulation by HIF-1 and MYC of metabolic enzymes and nutrient transporters.

## mTOR INTEGRATES NUTRIENT AVAILABILITY AND GROWTH FACTOR SIGNALING

As noted above, both nutrients and RTKs regulate metabolism. The atypical serine/threonine protein kinase mTOR integrates metabolic regulation by growth factors and nutrients. The identification of mTOR began in the early 1990s in genetic screens in yeast that elucidated target of rapamycin (TOR1 and TOR2) as mediators of the toxic effects of rapamycin on yeast. Subsequently, biochemical approaches led to the discovery of mTOR. On entry into mammalian cells, rapamycin binds to FKBP12, creating a drug–receptor complex that binds mTOR. The mechanism by which the rapamcyin–FKBP12 interaction disrupts mTOR function is not fully understood. mTOR is a component of two large protein complexes known as mTOR Complex I (mTORC1) and mTOR Complex 2 (mTORC2). The two complexes share common proteins, but they also have distinct proteins within their complexes. Importantly, the two complexes participate in divergent biological functions.

In the presence of ample nutrients and growth factors, mTORC1 activation promotes anabolic pathways involved in protein synthesis and lipid synthesis and also stimulates glycolysis and mitochondrial metabolism (Fig. 10-2). mTORC1 suppression blocks these pathways and induces autophagy and lysosomal bio-genesis. A salient feature of mTORC1 is that it changes its activity in response to availability of amino acids, oxygen, and growth factors, as well as decreases in the energetic status of cells. The upstream regulator of mTORC1 is the hetero-dimer consisting of tuberous sclerosis 1 and 2 (TSC1 and TSC2), which functions as a GTPase-activating protein for the Ras homolog enriched in brain (RHEB) GTPase. The GTP-bound RHEB stimulates mTORC1 kinase activity, and the TSC1/2 complex converts RHEB into the inactive GDP-bound state. Thus, repressing TSC1/2 allows RHEB to stimulate mTORC1 kinase activity. There are many inputs that either repress or activate TSC1/TSC2 to control mTORC1 activity. Growth factor activation of AKT inhibits TSC1/TSC2, resulting in mTORC1 stimulation to promote anabolism. In contrast, when cells are under limited oxygen (hypoxia) or energetically stressed, mTORC1 is inhibited to block energy-consuming anabolic pathways and stimulate catabolic pathways, such as autophagy.

The availability of amino acids also regulates mTORC1 activity, but not through TSC1/TSC2. Instead, the abundance of amino acids, particularly leucine and argi-nine, through unknown mechanisms is sensed at the lysosome surface to activate RAG (Ras-related GTP binding) GTPases, which promote mTORC1 activation. It is important to note that when cells are energetically stressed or have diminished oxy-gen or amino acid availability, they inhibit mTORC1 activity, even in the presence of growth factor stimulation. Thus, growth factors cannot promote mTORC1-dependent anabolism in the absence of nutrients. In contrast to mTORC1, mTORC2 primarily responds to growth factors and is insensitive to nutrients. It is not clear how mTORC2 is activated, but recent evidence suggests that perhaps ribosomes bind mTORC2 in a PI3K-dependent manner to activate mTORC2. This activation results in phosphorylation of the AGC subfamily of kinases, notably phosphorylating AKT at Ser-473, which enhances the strength of AKT activation downstream from PI3K and increases the range of effective AKT substrates. Understanding the complexity of mTOR pathways is important for the development of drugs that target this pathway given that laboratory studies indicate that mTOR activation is critical for nutrient overload resulting in obesity, whereas pharmacologic or genetic reduction in mTOR activity in multiple model organisms extends life span.

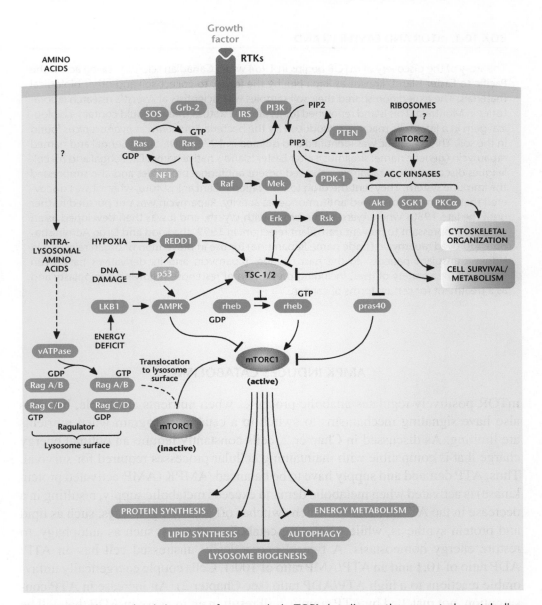

**Figure 10-2.** The mechanistic target of rapamycin (mTOR) signaling pathway controls metabolism and cell survival. mTORC1 stimulates multiple anabolic metabolic pathways (lipid and protein synthesis) in the presence of nutrients and growth factors. Conversely, limiting nutrient or growth factor availability promotes catabolic pathways (autophagy). mTORC2 promotes cell survival through activation of AKT. (Adapted from Laplante and Sabatini 2012.)

---

**BOX 10-1. mTOR AND EASTER ISLAND**

The story of the discovery of mTOR begins in 1964 when Canadian scientists sailed across the Pacific to Easter Island, known as Rapa Nui by the locals, to collect soil and other biological materials. The expedition shared their soil samples with scientists at Ayerst's research laboratories in Montreal. This island renowned for its famous statues (moai) would contain a biological gem in a lipophilic macrolide produced by the bacteria *Streptomyces hygroscopicus* found in the soil. This antibiotic was identified and isolated in 1972 by Dr. Suren Sehgal and named rapamycin (generic name: sirolimus), after Easter Island's native name. Dr. Sehgal and his colleagues discovered that rapamycin showed potent antifungal properties and also suppressed the immune system. They sent the drug to the National Cancer Institute, where it was discovered that rapamycin displayed antitumorigenic activity. Rapamycin was not pursued further until the late 1980s when Ayerst had merged with Wyeth, and it was then developed as an immunosuppressant to prevent transplant rejection. In 1999, the Food and Drug Administration approved rapamycin (trade name Rapamune) for use in prevention of organ rejection in kidney transplant patients. In the past decade, rapamycin and its derivatives have been approved for a variety of uses, including prevention of restenosis following angioplasty and as a treatment for certain forms of cancer.

## AMPK INDUCES CATABOLISM

mTOR positively regulates anabolic processes when nutrients are ample, but cells also have signaling mechanisms to switch to a catabolic program when nutrients are limiting. As discussed in Chapter 2, cells constantly require an optimal energy charge that is compatible with maintaining cellular processes required for survival. Thus, ATP demand and supply have to be balanced. AMPK (AMP-activated protein kinase) is activated when metabolic demand exceeds metabolic supply, resulting in a decrease in the ATP/ADP ratio, which switches off anabolic pathways, such as lipid and protein synthesis, while promoting catabolic pathways, such as autophagy, to restore energy homeostasis. A typical energetically unstressed cell has an ATP/ADP ratio of 10:1 and an ATP/AMP ratio of 100:1. Cells couple energetically unfavorable reactions to a high ATP/ADP ratio (see Chapter 2). An increase in ATP consumption, not matched by ATP supply, will result in an increase in ADP that will be rapidly converted to ATP by adenylate kinases, which catalyze the reversible reaction $2ADP \leftrightarrow ATP + AMP$. When this occurs, this reaction generates a drastic increase in AMP levels compared with changes in ATP or ADP.

The kinase AMPK responds to increases in AMP levels (Fig. 10-3). AMPK exists as a heterotrimer, consisting of a catalytic subunit ($\alpha$) and two regulatory subunits ($\beta$ and $\gamma$). The $\gamma$ subunit binds to AMP, ADP, and ATP to regulate the phosphorylation of Thr172 within the activation loop of the $\alpha$-subunit kinase domain, the key step in promoting AMPK activation. There are multiple mechanisms by which AMP promotes AMPK activation. First, AMP causes promotion of Thr172

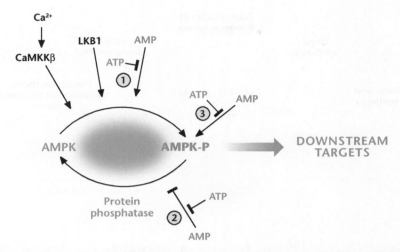

**Figure 10-3.** AMP and calcium activate AMP-activated protein kinase (AMPK). AMP displaces ATP on the AMPK-$\gamma$ subunit to (1) promote phosphorylation and (2) inhibit dephosphorylation of threonine-172, resulting in robust activation of AMPK. The constitutively active kinase LKB1 is required for phosphorylation of threonine-172. AMP further activates the phosphorylated AMPK (3). AMPK can also be activated by threonine-172 phosphorylation catalyzed by CaMKKβ, which also activates AMPK at threonine-172 through an increase in intracellular calcium, but independent of AMP and LKB1. (Adapted from Hardie and Alessi 2013.)

phosphorylation by a complex containing the tumor suppressor kinase liver kinase B1 (LKB1). Second, the effect of increased phosphorylation is amplified up to 10-fold further by allosteric activation by binding of AMP. Third, both ADP and AMP inhibit dephosphorylation of Thr172 by the phosphatases. However, AMP is about 10-fold more potent than ADP. ATP antagonizes all of these effects. AMPK can also be activated by Thr-172 phosphorylation catalyzed by CaMKKβ through a mechanism that requires an increase in intracellular calcium without any changes in adenine nucleotides.

Originally, AMPK was discovered to phosphorylate and inactivate the enzymes acetyl-CoA carboxylase (ACC) and 3-hydroxy-3-methylglutaryl-CoA reductase, which are key regulators of fatty acid and sterol biosynthesis, respectively. Subsequently, AMPK was discovered to phosphorylate multiple metabolic enzymes, as well as transcription factors and coactivators, to collectively diminish anabolic pathways that use NADPH and ATP while activating catabolic pathways that generate ATP (Fig. 10-4). For example, AMPK also promotes fatty acid oxidation in the mitochondria to increase ATP production while preventing de novo fatty acid synthesis, an ATP- and NADPH-consuming pathway. AMPK also suppresses the anabolic kinase mTORC1 activity by phosphorylating TSC2 and raptor, a regulatory subunit of mTORC1. The major catabolic program that AMPK engages in is the activation of autophagy, discussed in the next section.

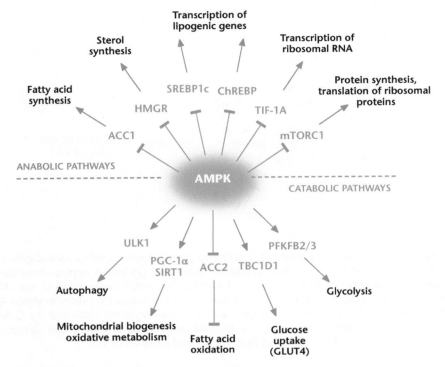

**Figure 10-4.** AMP-activated protein kinase (AMPK) regulates metabolism. AMPK positively and negatively regulates catabolic and anabolic pathways, respectively. (Adapted from Hardie and Alessi 2013.)

## AUTOPHAGY IS A CATABOLIC PROGRAM THAT MAINTAINS SURVIVAL

The late Christian de Duve (1974 Nobel Prize winner for the discovery of lyso-somes), in the 1960s, observed and coined the term "autophagy." Decades later, in the early 1990s, the molecular details underlying autophagy began to be eluci-dated largely through genetic approaches in yeast that analyzed autophagy-defective mutants. These studies identified multiple autophagy-related genes (ATGs), many of which are functionally conserved in mammals. Autophagy is thought to have evolved as a stress response to allow unicellular eukaryotic organisms to survive star-vation by generating amino acids needed to produce a subset of proteins required for adaptation to starvation and are catabolized through the TCA cycle to produce ATP for survival. Autophagy has been shown to be crucial for surviving starva-tion in multiple organisms, including yeast, flies, and worms. In mice, autophagy is up-regulated in most tissues, except the nervous tissues, and shortly after birth, when nutrient supply from the placenta is abruptly terminated. Autophagy-deficient mice display normal amino acid levels at birth, but die within hours after birth because of their inability to maintain the amino acid pool. Interestingly, induction

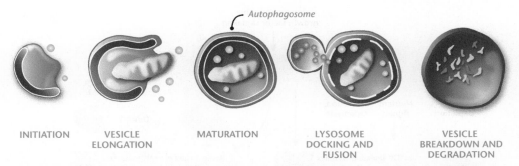

*Autophagosome*

| INITIATION | VESICLE ELONGATION | MATURATION | LYSOSOME DOCKING AND FUSION | VESICLE BREAKDOWN AND DEGRADATION |

**Figure 10-5.** Overview of autophagy. The autophagy pathway has multiple steps, including an autophagosome that contains damaged organelles that are degraded in the final stage. This ensures organelle quality control and can generate metabolites for ATP production during nutrient-limiting conditions. (Adapted, with permission, from Choi et al. 2013, © Massachusetts Medical Society.)

of low levels of autophagy is thought to promote longevity by clearing damaged organelles and aberrant proteins that accumulate over time.

There are three types of autophagy: macroautophagy, microautophagy, and chaperone-mediated autophagy (CMA) (Fig. 10-5). Macroautophagy occurs when large regions of the cytosol containing organelles are sequestered into a double-membrane vesicle called the autophagosome, which fuses with lysosomes for degradation; this generates metabolites, such as amino acids, which can be used to generate ATP for survival. In contrast, in microautophagy, the lysosomal membrane invaginates to sequester small regions of the cytosol, which are internalized into the lysosomal lumen as single-membrane vesicles. In CMA, substrate proteins are specifically recognized by a cytosolic chaperone that transports them to the surface of the lysosome where they translocate by receptor-mediated mechanisms to the lysosomal lumen for degradation. It is important to note that autophagy occurs at a basal level in the presence of growth factors and ample nutrients as it is involved in degradation of damaged organelles and proteins. Nutrient and/or growth factor deprivation that limits nutrient uptake dramatically induce macroautophagy.

The macroautophagy pathway can be broken down into several discrete phases, including initiation involving the formation of a preautophagosomal structure leading to an isolation membrane followed by vesicle expansion and autophagosome maturation. Subsequently, the autophagosome fuses with the lysosome and the autophagosomal contents are degraded by lysosomal acid hydrolases generating metabolites, such as amino acids, which can be used as a fuel source to maintain an optimal energy charge for survival. These molecular events are executed by homologs of products of the ATGs (*Atg*) originally identified in yeast. Autophagosomal membrane initiation requires translocation of a complex consisting of UNC-51-like kinase 1 (ULK1), ATG13, ATG101, and FIP200 from the cytosol to certain domains of the endoplasmic reticulum, resulting in the recruitment of the Beclin 1 interacting complex (consisting of Beclin 1, the class III phosphatidylinositol 3-kinase [VPS34],

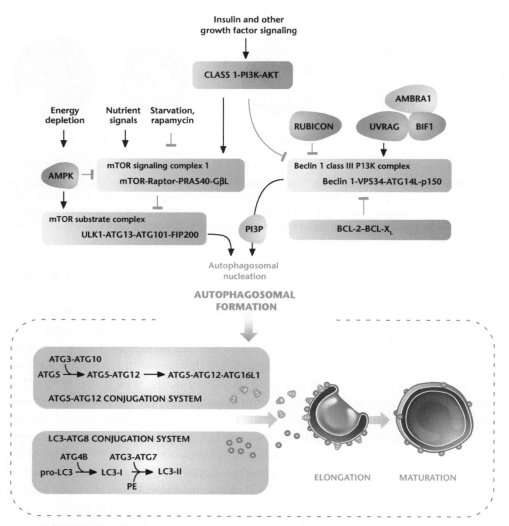

**Figure 10-6.** Regulators of autophagy. Growth factors and nutrients through PI3K/AKT and mTORC1, respectively, negatively regulate autophagy. AMP-activated protein kinase (AMPK) positively regulates autophagy. Autophagy is also regulated by the Beclin 1 interacting complex. Autophagosomal elongation requires the ATG5-ATG12 and LC3-ATG8 ubiquitin-like conjugation systems. In mammals, the conversion of LC3-I (free-form) to LC3-II (phosphoethanolamine-conjugated form) is used as a marker of autophagy. (Adapted, with permission, from Choi et al. 2013, © Massachusetts Medical Society.)

and ATG14 L) to generate phosphatidylinositol-3-phosphate (PIP3), the key initiating step (Fig. 10-6). Autophagosomal expansion requires two ubiquitin-like conjugation systems, the microtubule-associated protein light chain 3 (LC3–ATG8) and the ATG5–ATG12 complex. Autophagosome formation is indicated by the conversion of a cytosolic truncated form of LC3 (LC3-I) to its autophagosomal membrane-associated, phosphatidylethanolamine-conjugated form (LC3-II).

Autophagy has to be tightly controlled so that it is not inadvertently stimulated when intracellular nutrient availability is ample and cells are in an anabolic mode. The kinases mTORC1 and AMPK are two dominant regulators of autophagy. mTORC1 is maintained in an active conformation to concomitantly stimulate an anabolic program compatible with cellular growth while suppressing the catabolic program of autophagy when intracellular amino acids and other nutrients are abundant. mTORC1 phosphorylates and negatively regulates ULK1 to suppress autophagy when nutrients are limiting, especially amino acids. Nutrient limitation also activates AMPK (as discussed above), which inhibits mTORC1, and also directly phosphorylates ULK1 to initiate autophagy. Thus, ULK1 integrates mTORC1 and AMPK signaling to regulate autophagy.

## OXYGEN AND GLUCOSE REGULATE TRANSCRIPTIONAL NETWORKS

Two nutrient changes that cells respond to are those in environmental oxygen and glucose levels. Oxygen homeostasis is essential for the life and health of metazoans. Thus, multiple mechanisms evolved for adaptation to decreases in oxygen levels (i.e., hypoxia). These organismal adaptive responses include increased ventilation and constriction of the pulmonary artery, angiogenesis (formation of new blood vessels), and erythropoiesis (generation of red blood cells), as well as a cellular transcriptional program that promotes glycolysis. It makes sense that as oxygen levels are diminishing there would be enhanced glycolysis coupled with angiogenesis to bring more blood flow, as well as stimulation of erythropoiesis to increase oxygen-carrying capacity.

At the cellular level, oxygen levels starting at 5% begin to noticeably activate the hypoxia-inducible transcription factors (HIFs) (Fig. 10-7). HIFs bind to hypoxia-response elements in the promoter/enhancer regions of a large number of target genes, including all the glycolytic enzymes, erythropoietin to simulate erythropoiesis, and proangiogenic factor vascular endothelial growth factor. HIFs are a heterodimer consisting of two basic helix–loop–helix/PAS proteins, HIF-1α or HIF-2α and HIF-1β. The HIF-1β subunit is constitutively present in the cell, but the HIFα subunit is only present during hypoxia. During normoxia, HIF-1α is polyubiquitinated and targeted for degradation by an E3 ubiquitin ligase, which contains the specificity factor, the von Hippel–Lindau tumor suppressor (VHL). The binding of VHL to HIFα is dependent on the hydroxylation of proline residues within the oxygen-dependent degradation domain (ODD) of HIFα. A family of α-ketoglutarate-dependent dioxygenases, termed prolyl hydroxylases 1, 2, and 3 (PHDs 1–3), catalyze this hydroxylation event. Genetic studies have revealed that PHD2 is the primary enzyme responsible for hydroxylation of the HIFα subunit. The activity of PHD2 is inhibited under hypoxic conditions, allowing the accumulation of HIFα protein and the subsequent binding to HIFβ subunit to induce the transcriptional response. HIF1

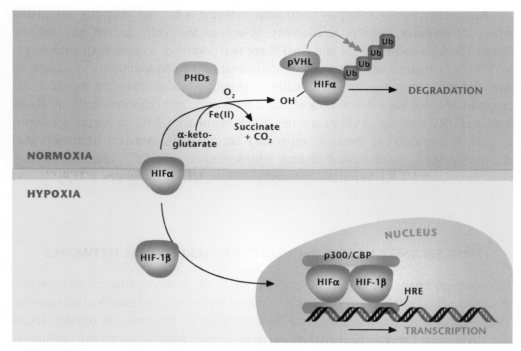

**Figure 10-7.** Oxygen levels control stabilization of HIFs. HIFs are a heterodimer, consisting of a constitutively stable HIF-1β and an oxygen-sensitive HIFα subunit. The HIFα subunit is hydroxylated at proline residues in the ODD by PHDs under normoxia. In addition to oxygen, the PHDs require iron ($Fe^{2+}$) and α-ketoglutrate as substrates for hydroxylation. The von Hippel–Lindau protein (pVHL) recognizes hydroxylated proline residues and targets the HIFα subunit for rapid degradation by the proteasome. Hypoxia diminishes hydroxylation of HIFα to prevent its degradation and promote dimerization with HIF-1β to bind specific hypoxia response elements (HRE) in the promoter regions of its target genes. (Adapted, with permission, from Balligand et al. 2009, © The American Physiological Society.)

and -2 have distinct and overlapping gene targets. HIFs are activated by other stimuli, notably ROS, and have been implicated in pathological conditions including pulmonary hypertension, inflammation, and cancer.

An important adaptation of hypoxia is to reduce metabolism as measured by a decrease in the mitochondrial oxygen consumption of cells (i.e., the respiratory rate) occurring at oxygen levels (1%–3% $O_2$) that are well above the threshold in which oxygen becomes limiting (<0.3% $O_2$) to cytochrome $c$ oxidase (complex IV), a phenomenon referred to as oxygen conformance. Complex IV is the main enzyme in the electron transport chain (ETC) that uses oxygen and couples it to the generation of ATP (see Chapter 4). ATP, ADP, and AMP levels do not change in 1%–3% $O_2$ levels. Nevertheless, AMPK is activated within minutes through CaMKKβ at these oxygen levels to repress mTORC1, thus, shifting from anabolism to catabolism. Within minutes, hypoxia also decreases the activity of the plasma membrane Na/K-ATPase, a major ATP consumer. Longer exposure to hypoxia

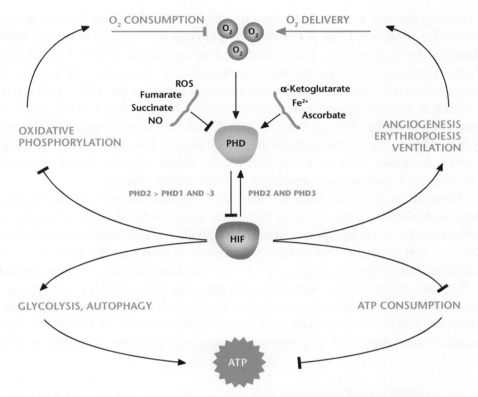

**Figure 10-8.** HIFs regulate adaptation to hypoxia. Multiple inputs, including hypoxia, ROS, TCA metabolites (succinate and fumarate), and NO, inhibit prolyl hydroxlases (PHDs) to activate hypoxia inducible factors (HIFs). At the organismal level, HIFs stimulate angiogenesis, erythropoiesis, and ventilation to increase oxygen delivery. At the cellular level, HIFs diminish oxidative phosphorylation and increase glycolysis. HIFs also suppress ATP-consuming generation to preserve a high ATP/ADP ratio in cells. (Adapted from Kaelin and Ratcliffe 2008.)

results in HIF-1-mediated repression of mTORC1 and other anabolic processes while stimulating glycolysis and autophagy (Fig. 10-8).

A perplexing observation under hypoxia is that cells display an increase in ROS generated from the mitochondrial ETC. The details underlying this phenomenon are not fully understood. However, suppression of mitochondrial ROS can diminish many of the effects of hypoxia, such as HIF and AMPK activation. Thus, mitochondrial ROS serve as signaling molecules that couple lowering of oxygen levels to multiple biological effects. A long-term consequence of hypoxia would be a marked state of high ROS, which could accumulate to levels that would cause oxidative damage to cells. Cells have adapted strategies to counteract the buildup of ROS during hypoxia. In particular, HIF-1 decreases pyruvate conversion to acetyl-CoA through induction of lactate dehydrogenase A (LDH-A) and pyruvate dehydrogenase kinase 1 (PDK1). PDK1 phosphorylates and inactivates the catalytic subunit of pyruvate dehydrogenase (PDH). An increase in PDK1 protein levels decreases PDH activity, thereby preventing the conversion of pyruvate to acetyl-CoA, while also driving pyruvate conversion to lactate. LDH-A converts the pyruvate to lactate by using the NADH generated from glycolysis. The coordinated up-regulation of PDK1 and LDH-A diverts pyruvate from fueling the mitochondria to generation of lactate. The decreased acetyl-CoA levels in the mitochondria diminish the TCA cycle activity, resulting in reduced generation of mitochondrial NADH and $FADH_2$ levels and reduced electron flux through the ETC. HIF-1 also decreases complex I and IV activity by inducing genes that regulate these complexes. Indeed, loss of HIF1 under hypoxia increases ROS to levels that induce death. Overall, HIF-1 diminishes ETC activity to prevent overproduction of ROS during hypoxia.

Mammalian cells also have mechanisms to monitor changes in glucose levels and couple them to transcriptional responses. Intracellular glucose is rapidly converted into glucose 6-phosphate, which, through unidentified mechanisms, activates MondoA and ChREBP that in turn heterodimerize with the Mlx protein to make functional transcriptional factors (Fig. 10-9). The MondoA/Mlx complex localizes to the outer mitochondrial membrane, whereas the ChREBP/Mlx is localized in the cytosol under low-glucose conditions. In response to an increase in glucose 6-phosphate levels, MondoA/Mlx and ChREBP/Mlx complexes translocate to the nucleus where they bind to carbohydrate response elements (ChoRE), resulting

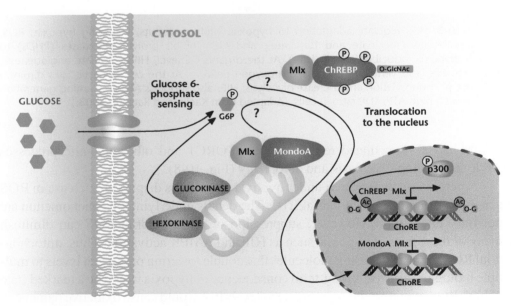

**Figure 10-9.** Mondo transcription factors respond to glucose flux. Mondo transcription factors ChREBP and MondoA, with their common binding partner Mlx, are stimulated through unknown mechanisms to an increase in glucose 6-phosphate levels. Subsequently, MondoA/Mlx and ChREBP/Mlx complexes translocate to the nucleus where they bind to ChoRE, resulting in activation of genes that encode enzymes in glycolysis and lipid synthesis. (Adapted from Havula and Hietakangas 2012, with permission from Elsevier.)

in activation of genes that encode enzymes in glycolysis and lipid synthesis. The initial identification of ChoRE allowed the purification of ChREBP (carbohydrate response element binding protein). ChREBP also undergoes various PTMs, notably *O*-GlcNAcylation, for maximal transcriptional activation.

A downstream production of glucose 6-phosphate is fructose 6-phosphate, which can funnel into the hexosamine pathway to generate *O*-GlcNAcylation. ChREBP is detectable in most tissues, but is at its highest levels in liver. The ChREBP/Mlx complex redirects glucose metabolism to support lipogenesis by inducing a number of key genes involved in fatty acid synthesis, *including* ACC, fatty acid synthase, and stearoyl-CoA desaturase 1. The ChREBP/Mlx complex also targets induction of the liver pyruvate kinase gene. In the fasting state, hepatic glycolysis and lipogenesis are suppressed, whereas gluconeogenesis and fatty acid oxidation are enhanced (see Chapter 6). This shift from anabolism to catabolism is regulated by hormones glucagon and epinephrine, which increase cAMP concentration and activate cAMP-activated protein kinase, which phosphorylates and inactivates ChREBP. Similarly, intracellular accumulation of AMP inhibits ChREBP through AMPK activation. The Mondo/Mlx complex is also detectable in most tissues, but it is strongly expressed in skeletal muscle. The Mondo/Mlx complex transcriptionally induces many glycolytic enzymes, including the rate-limiting enzymes. Interestingly, growth factor signaling promotes protein ChREBP expression. In proliferating cells, such as cancer cells, the ChREBP/Mlx complex promotes genes involved in de novo lipid and nucleotide metabolism. Interestingly, growth factor signaling promotes protein ChREBP expression. Thus, growth factor signaling coordinates two key aspects: (1) stimulation of glucose uptake and glycolysis to produce glucose 6-phosphate and (2) the expression of ChREBP.

## INTRACELLULAR METABOLITES CONTROL SIGNALING

In addition to control of signaling by growth factor signaling and nutrients, intracellular metabolites can exert feedback control on signaling through PTMs of proteins, such as glycosylation, acetylation, methylation, and prenylation (Fig. 10-10). These covalent modifications by metabolites change activity, localization, or stability of target proteins. The metabolic fluxes through different pathways required to produce these PTM substrates fluctuate in response to intracellular metabolite availability, as well as different cellular states—differentiation versus proliferation.

A key metabolite that governs many of these PTMs is acetyl-CoA. As discussed in Chapter 7, acetyl-CoA is a precursor for fatty acid and cholesterol synthesis pathways, which provide substrates for palmitoylation, myristoylation, and prenylation. Acetyl-CoA also is a substrate for protein acetylation, notably on lysine residues. Acetylation is controlled by the collective activity of acetyltransferases and deacetylases. Acetyl-CoA, which is produced in the mitochondria, condenses with

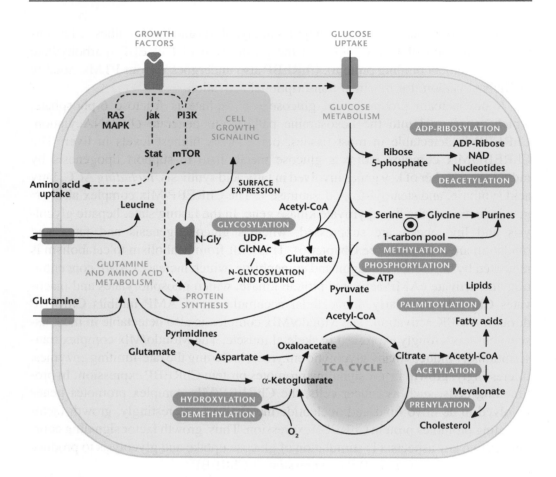

**Figure 10-10.** Metabolic pathways regulate signaling pathways. Metabolic pathways generate substrates that are used as posttranslational modifications (PTMs) to control signal transduction. Notable PTMs highlighted include glycosylation, prenylation, palmitoylation, hydroxylation, methylation/demethylation, phoshorylation, acetylation/deacyetylation, and ADP-ribosylation. (Adapted, with permission, from Metallo and Vander Heiden 2010.)

oxaloacetate to generate citrate, which can be exported into the cytosol. Subsequently, the enzyme ATP citrate lyase (ACLY) converts citrate into oxaloacetate and acetyl-CoA. Thus, mitochondrial oxidative metabolism is a major source of acetyl-CoA used for protein acetylation. The other source is acetate produced by the bacteria in the gut. Acetate is converted into acetyl-CoA in the cytosol by ACECS1. Both ACL and ACECS1 are found in the nucleus and cytosol of mammalian cells; ACECS2 is the mitochondrial isoform. Not all cells have access to sufficient quantities of acetate; thus, ACLY-dependent reaction is the major source of acetyl-CoA in mammalian cells. One challenge in the field is to quantitatively measure acetyl-CoA levels under different cellular conditions in the nucleus, cytosol, and mitochondria.

The availability of acetyl-CoA in the cytosol from mitochondria increases in conditions of anabolic states, cellular proliferation, and differentiation. In contrast, most of the mitochondrial acetyl-CoA generated during the quiescent state is used to generate ATP and not exported into the cytosol. A consequence of this increase in acetyl-CoA levels is to acetylate histones at specific lysine residues, loosening the interaction between the negatively charged DNA and the histone and opening the chromatin structure to induce gene expression that promotes proliferation or differentiation. If signals for proliferation or differentiation exist, but metabolism is not induced to generate sufficient acetyl-CoA, then the failure to acetylate histones provides a failsafe mechanism to halt the gene induction that promotes proliferation or differentiation. If the metabolic state of the cells were not linked to induction of genes, then cells would begin to undergo proliferation and differentiation without "realizing" whether they have enough biosynthetic and bioenergetic capacity to undertake a metabolically demanding process, such as proliferation or differentiation. Linking metabolism to gene induction is a conserved mechanism. When yeast cells enter a growth phase, transcription of cell growth genes is induced because of the increased histone acetylation that occurs as a result of increased acetyl-CoA production.

Acetylation also occurs extensively on lysine residues of many metabolic enzymes, including, but not limited to, those involved in glycolysis, fatty acid oxidation, and the TCA cycle. Typically, acetylation of these enzymes diminishes their activity and slows down metabolism. This is essentially a feedback of excess intracellular metabolite production readout by high levels of acetyl-CoA to not accelerate metabolism further because cells already are engaged in robust metabolism. In contrast, if metabolism is in a catabolic state, there is no need to further dampen metabolic enzymes by acetylation because energy generation is paramount under these conditions. Not all protein acetylation is sensitive to changes in acetyl-CoA concentration. For example, changes in acetyl-CoA levels that alter histone acetylation do not elicit changes in tubulin acetylation. Thus, acetyltransferases are likely to have different $K_m$ values for levels of acetyl-CoA (Fig. 10-11).

Deacetylase enzymes are another major input in controlling acetylation of proteins. These enzymes catalyze deacetylation in a $NAD^+$-independent or -dependent manner. Of these enzymes, the family of $NAD^+$-dependent deacetylases is the best characterized. These enzymes use $NAD^+$ as a substrate to remove the acetyl group from the protein, yielding O-acetyl-ADP-ribose plus nicotinamide (NAM) as a product. The levels of $NAD^+$ are tightly regulated through either the salvage or de novo pathway. In the salvage pathway, nicotinamide phosphoribosyltransferase (NAMPT) converts NAM to NMN (nicotinamide mononucleotide) and, subsequently, NMNAT (nicotinamide nucleotide adenylyltransferase 1) converts NMN to $NAD^+$ (Fig. 10-12). In the de novo pathway, tryptophan is converted in multiple steps to produce nicotinic acid mononucleotide and, subsequently, $NAD^+$.

The seven members of the sirtuin (SIRT1–7) family are localized to different compartments of the cells and, thus, have different functions. Mitochondrial and

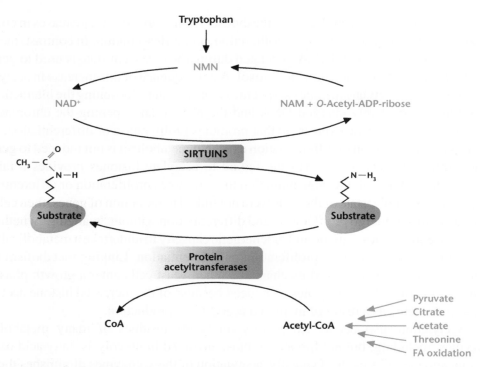

**Figure 10-11.** Metabolism regulates acetylation and deacetylation. Acetyl-CoA provides the acetyl group for acetylation reactions catalyzed by protein acetyltransferases. The Sirtuin family of deacetylases uses NAD$^+$ to remove acetyl group from proteins. (Adapted from Kaelin and McKnight 2013.)

cytoplasmic sirtuins control the acetylation status of metabolic enzymes, whereas nuclear sirtuins control acetylation of factors that control transcription of genes. Broadly, these deacetylases help organisms adapt to conditions when nutrient availability is diminished. Sirtuin's activity also declines with aging, which is linked to age-related pathologies. As discussed above, acetylation decreases activity of metabolic enzymes, but under nutrient-limiting conditions, cells deacetylate these enzymes to allow them to maximize their ability to generate glycolyltic and TCA-cycle metabolites necessary for energy generation. Also, under nutrient-limiting conditions, AMPK phosphorylates SIRT1 to increase its activity causing deacetylation of certain proteins in the nucleus, to allow activation of genes required for metabolic adaptation. Indeed, caloric restriction increases expression of genes that control NAD$^+$ levels and sirtuins that collectively activate enzymes and genes for metabolic adaptation. Current efforts are examining mechanisms to boost sirtuin activity to ameliorate diseases. For example, the supplementation of NMN to boost NAD$^+$ levels has been shown to prevent diet- and age-induced type II diabetes in mice.

Methylation of proteins (notably histones) and DNA is another PTM regulated by metabolism. Methylation is controlled by the collective activity of methyltransferases and demethylases (Fig. 10-13). *S*-adenosylmethionine (SAM) is the methyl

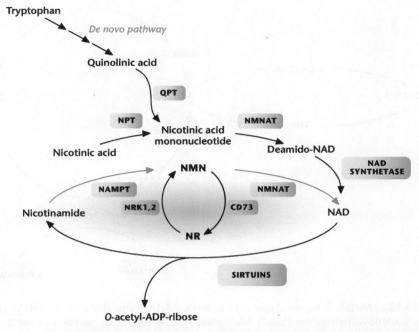

**Figure 10-12.** Metabolic pathways that generate NAD⁺ in mammals. Tryptophan, nicotinic acid, nicotinamide (NAM), or nicotinamide riboside (NR) are all NAD⁺ precursors. Enzymes in the pathways are quinolinate phosphoribosyltransferase (QPT), NMNAT, nicotinamide riboside kinase 1 and (NRK1,2), NAMPT, nicotinic acid phosphoribosyltransferase (NPT), and ecto-5′-nucleotidase (CD73). (Adapted from Imai and Guarente 2014, with permission from Elsevier.)

donor used by most methyltransferases, including those that methylate histones and DNA. SAM is derived from methinone and becomes *S*-adenosylhomocysteine (SAH) on donating its methyl group. SAH is converted into homocysteine, which is then converted back to a vitamin B12-dependent reaction that uses carbons derived from folate or choline. Thus, methyltransferases are dependent on a supply of methionine or carbon flux into the folate pool.

An instructive example of metabolism driving epigenetic changes comes from the observation that the highly proliferative mouse embryonic stem cells (ESCs) require threonine as an amino acid to maintain their pluripotent state. Mouse ESCs express robust levels of the threonine dehydrogenase enzyme, which catabolizes threonine into 2-amino-3-ketobutyrate. Subsequently, 2-amino-3-ketobutyrate ligase using coenzyme A yields glycine and acetyl-CoA. Glycine, ultimately, can generate SAM required for methylation.

Metabolites also control demethylases. Lysine-specific demethylase 1 (LSD1) is dependent on FAD reduction to FADH$_2$ (Fig. 10-14). FAD is produced in the mitochondria, and changes in FAD availability to LSD1 might regulate its activity. The family of demethylases that contain the Jumonji C (JmjC) domain are members

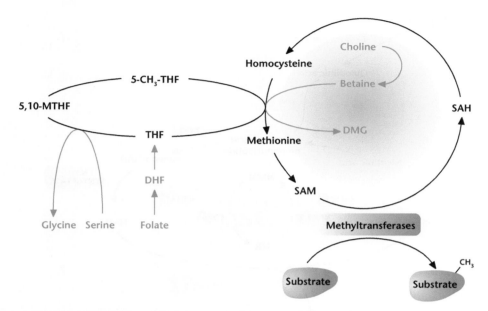

**Figure 10-13.** Metabolism regulates methylation. Methionine donates a methyl group to generate *S*-adenosylmethionine (SAM). Methyltransferases use SAM, resulting in the generation of *S*-adenosylhomocysteine (SAH), which is converted to homocysteine. Carbons derived from either choline- or folate-dependent reactions convert homocysteine back to methionine. DHF, dihydrofolate; THF, tetrahydrofolate; 5,10-MTHF, 5,10-methylene THF; $CH_3$, methyl. (Adapted from Kaelin and McKnight 2013.)

of the $\alpha$-ketoglutarate and $Fe^{2+}$-dependent dioxygenase family. Similar to PHDs discussed above, these proteins consume oxygen, convert $\alpha$-ketoglutrate to succinate, and require ascorbate to regenerate the $Fe^{2+}$ cofactor as part of their enzymatic activity. It is conceivable that changes in any of these metabolites would affect JmjC demethylase activity. Going forward, it will be important to determine whether any metabolites are rate limiting for methylation or demethylation reactions under physiological or pathophysiological conditions. The exception is oxygen, which decreases under physiological conditions, such as ascending to altitude, or under pathophysiological conditions, such as ischemic diseases or lung diseases like chronic obstructive pulmonary disease. Indeed, there is burgeoning evidence that decreases in oxygen levels influence the epigenetic state of cells and organisms by regulating oxygen-dependent demethylases.

This chapter started with a discussion of the importance of growth factors in nutrient uptake through activation of signaling pathways. A consequence of nutrient uptake, in particular, glucose and glutamine, is that glucose and glutamine allow glycosylation of growth factors, which is critical for the proper folding of the growth factor receptors and their localization to the cell surface. As discussed in Figure 10 of Chapter 6, glycosylation is dependent on the glucose metabolism through the hexosamine biosynthetic pathway, a subsidiary pathway of glycolysis, to generate

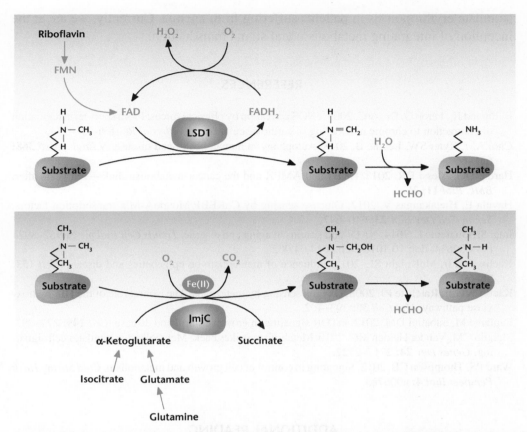

**Figure 10-14.** Metabolism regulates demethylation. Lysine-specific demethylases (LSD) couple oxidation of methyl groups in histones to reduction of FAD to $FADH_2$, which spontaneously causes demethylation of histone and produces formaldehyde (HCHO). JmjC demethylases use α-ketoglutarate, Fe(II), and oxygen to hydroxylate methylated histones resulting in their demethylation.

UDP-*N*-acetylglucosamine (UDP-GlcNAc). Glutamine serves as the nitrogen donor in the production of UDP-GlcNAc. Thus, glycosylation of growth factor receptors is an example of feedback of metabolism on growth factor signaling.

It is evident that the large signal transduction and metabolic pathway diagrams that many of us have plastered on our laboratory walls are interconnected. It is essential to appreciate the integration of signal transduction and metabolism because many of the current escalations in pathologies are induced by changes in our diet and habits (e.g., lack of exercise). Metabolism and signal transduction pathways are poised to detect changes in diet and habits, which ultimately modulate transcriptional networks to alter the organismal phenotype. The understanding of cross talk between metabolism and signal transduction will provide possible targets of therapies. An example is the current effort to make drugs that inhibit PHDs to activate HIFs to stimulate erythropoiesis in patients suffering from anemia. Currently, we are at the inception of integrating metabolism and signal transduction.

## REFERENCES

Balligand JL, Feron O, Dessy C. 2009. eNOS activation by physical forces: From short-term regulation of contraction to chronic remodeling of cardiovascular tissues. *Physiol Rev* **89:** 481–534.

Choi AM, Ryter SW, Levine B. 2013. Autophagy in human health and disease. *N Engl J Med* **368:** 651–662.

Hardie DG, Alessi DR. 2013. LKB1 and AMPK and the cancer-metabolism link—Ten years after. *BMC Biol* **11:** 36.

Havula E, Hietakangas V. 2012. Glucose sensing by ChREBP/MondoA-Mlx transcription factors. *Semin Cell Dev Biol* **23:** 640–647.

Imai SI, Guarente L. 2014. NAD$^+$ and sirtuins in aging and disease. *Trends Cell Biol* pii: S0962–8924 (14)000634. doi: 10.1016/j.tcb.2014.04.002.

Kaelin WG Jr, McKnight SL. 2013. Influence of metabolism on epigenetics and disease. *Cell* **153:** 56–69.

Kaelin WG Jr, Ratcliffe PJ. 2008. Oxygen sensing by metazoans: The central role of the HIF hydroxylase pathway. *Mol Cell* **30:** 393–402.

Laplante M, Sabatini DM. 2012. mTOR signaling in growth control and disease. *Cell* **149:** 274–293.

Metallo CM, Vander Heiden MG. 2010. Metabolism strikes back: Metabolic flux regulates cell signaling. *Genes Dev* **24:** 2717–2722.

Ward PS, Thompson CB. 2012. Signaling in control of cell growth and metabolism. *Cold Spring Harb Perspect Biol* **4:** a006783.

## ADDITIONAL READING

Cai L, Tu BP. 2012. Driving the cell cycle through metabolism. *Annu Rev Cell Dev Biol* **28:** 59–87.

Gowans GJ, Hardie DG. 2014. AMPK: A cellular energy sensor primarily regulated by AMP. *Biochem Soc Trans* **42:** 71–75.

Manning BD, Cantley LC. 2007. AKT/PKB signaling: Navigating downstream. *Cell* **129:** 1261–1274.

Mihaylova MM, Shaw RJ. 2011. The AMPK signalling pathway coordinates cell growth, autophagy and metabolism. *Nat Cell Biol* **13:** 1016–1023.

Sehgal SN. 2003. Sirolimus: Its discovery, biological properties, and mechanism of action. *Transplant Proc* **35** Suppl 3: 7S–14S.

Semenza GL. 2012. Hypoxia-inducible factors in physiology and medicine. *Cell* **148:** 399–408.

Thompson CB. 2011. Rethinking the regulation of cellular metabolism. *Cold Spring Harb Symp Quant Biol* **76:** 23–29.

Wellen KE, Thompson CB. 2012. A two-way street: Reciprocal regulation of metabolism and signalling. *Nat Rev Mol Cell Biol* **13:** 270–276.

Xiong Y, Guan KL. 2012. Mechanistic insights into the regulation of metabolic enzymes by acetylation. *J Cell Biol* **198:** 155–164.

# 11

# Metabolism of Proliferating Cells

A N INSTRUCTIVE EXAMPLE OF THE INTEGRATION OF multiple metabolic pathways covered in this book is to examine the metabolic needs of proliferating cells (e.g., T cells, stem cells, and cancer cells). A distinguishing feature of proliferating cells compared with nonproliferating cells (e.g., quiescent or differentiated cells) is the massive anabolism that proliferating cells undergo when they double their total biomass to subsequently divide into two daughter cells. Cell metabolism is reprogrammed to increase the uptake of nutrients that feed metabolic pathways to ultimately supply carbon, nitrogen, ATP, and NADPH for production of lipids, proteins, and nucleotides needed to build two daughter cells (Fig. 11-1). ATP and NADPH are necessary to drive many of the thermodynamically unfavorable anabolic

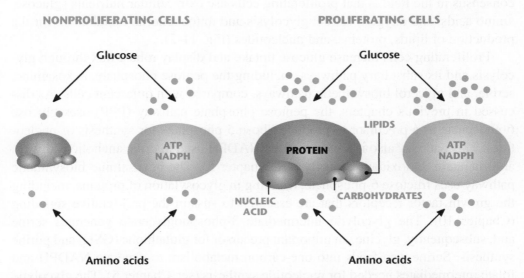

**Figure 11-1.** Proliferating versus nonproliferating cells have different metabolic needs. Nonproliferating cells do not have an excessive need to conduct anabolic functions, thus they catabolize their nutrients to generate ATP and NADPH for housekeeping functions. In contrast, proliferating cells engage in a massive anabolic program to generate lipids, proteins, and nucleotides. (Adapted from Vander Heiden 2011, by permission from Macmillan Publishers Ltd.)

reactions. ATP and NADPH also maintain housekeeping functions, such as maintenance of ion gradients across membranes and antioxidant capacity, respectively. In contrast, nonproliferating cells do not have an excessive need to conduct anabolic functions and catabolize their nutrients to generate ATP and NADPH for housekeeping functions (Fig. 11-1). This chapter focuses on the metabolism of proliferating cells with special attention on T- and cancer-cell proliferation as examples of normal- and malignant-cell proliferation, respectively.

## GLYCOLYSIS AND MITOCHONDRIAL METABOLISM ARE CENTRAL PATHWAYS SUPPORTING CELL PROLIFERATION

In the 1920s, Otto Warburg initially recognized that copious amounts of lactate are generated in rapidly proliferating ascites tumors (see Chapter 3). This phenomenon, termed the Warburg effect or aerobic glycolysis, has since been observed across several tumor types and proliferating T and embryonic stem cells. Many studies, including Warburg's study, conclude that proliferating cells do not engage robustly in mitochondrial metabolism and that aerobic glycolysis is the only major feature of proliferating cells' metabolic phenotype. However, as discussed in Chapter 4, mitochondria generate metabolites required for lipids, proteins, and nucleic acids. In fact, most cancer cells have functional mitochondrial oxidative metabolism. The current consensus in the field is that proliferating cells use extracellular nutrients (glucose, amino acids, and oxygen) to fuel glycolysis and mitochondrial metabolism for the production of lipids, proteins, and nucleotides (Fig. 11-2).

Proliferating cells increase glucose uptake and display robust flux through glycolysis and its subsidiary pathways, including the pentose phosphate, hexosamine, serine, and glycerol biosynthetic pathways, compared with quiescent cells. As discussed in previous chapters, the pentose phosphate pathway (PPP) uses glucose 6-phosphate as a precursor to produce ribose 5-phosphate for synthesis of nucleotides. This pathway also is a source of the NADPH used to drive anabolic pathways and maintain antioxidant capacity (see Chapter 5). The hexosamine biosynthetic pathway uses fructose 6-phosphate, resulting in glycosylation of proteins, including the growth factor receptors that are essential to sustain the proliferative signaling (Chapter 10). The glycolytic intermediate 3-phosphoglycerate generates serine and, subsequently, glycine, an important precursor for glutathione (GSH) and purine synthesis. Serine also feeds into one-carbon metabolism to generate NADPH and folate intermediates needed for nucleotide synthesis (see Chapter 5). The glycolytic intermediate glyceraldehyde 3-phosphate generates glycerol 3-phosphate, which is used to produce lipids.

The increase in glycolytic flux requires a constant supply of $NAD^+$, which is essential for the conversion of glyceraldehyde 3-phosphate to 1,3-bisphosphoglycerate by glyceraldehyde 3-phosphate dehydrogenase (step 6 of glycolysis). In this

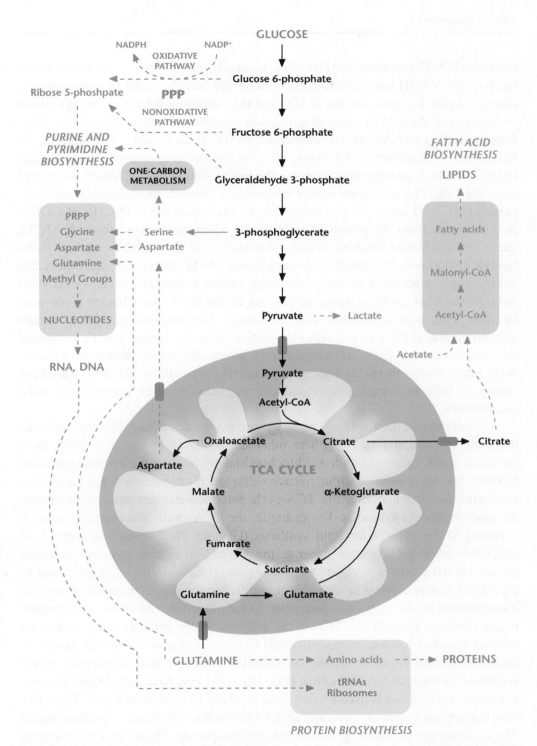

**Figure 11-2.** Proliferating cells require glycolysis and mitochondrial metabolism. Proliferating cells use extracellular nutrients (glucose, amino acids, and oxygen) to fuel glycolysis and its subsidiary pathways, including pentose phosphate pathway (PPP) and one-carbon metabolism, as well as mitochondrial metabolism for the production of lipids, proteins, and nucleotides. (Adapted from Deberardinis et al. 2008, with permission from Elsevier.)

reaction, $NAD^+$ becomes NADH. The regeneration of $NAD^+$ can occur either by shuttling NADH into mitochondria to allow the mitochondrial electron transport chain complex I to generate the $NAD^+$ and then shuttling it back to the cytoplasm or converting the end product of glycolysis pyruvate to lactate by lactate dehydrogenase A (LDH-A), which regenerates $NAD^+$ from NADH. The shuttling of NADH into mitochondria is a slower process. Proliferating cells express abundant levels of LDH-A protein, thus, favoring the production of lactate, which is secreted from the cell. The release or uptake of lactate is through monocarboxylate transporters (MCTs) 1 and 4. At physiological pH, the lactate ion ($CH_3CH(OH)COO^-$) is dissociated from its proton $H^+$. However, when transported through MCTs, the proton and lactate anion are associated as lactic acid. Thus, MCTs mediate membrane transport with 1:1 coupling between lactate and $H^+$ fluxes. The secreted lactate is not simply a wasteful product containing carbon molecules. It can be recycled through the Cori cycle to generate glucose in the liver (see Chapter 6) or used by other cells, such as neurons, as a fuel source. The reaction catalyzed by lactate dehydrogenase (LDH) is a reversible reaction whereby lactate can be converted into pyruvate, and pyruvate subsequently enters the mitochondria to generate ATP. Interestingly, lactate and accompanying $H^+$ (intracellular pH drop) have been shown to influence angiogenesis and cell migration through poorly understood mechanisms.

Glycolysis and its subsidiary pathways are not sufficient to fulfill the metabolic demands of proliferating cells. The other metabolic pathways that fulfill these demands are in the mitochondria. Mitochondrial one-carbon metabolism generates NADPH for redox balance within mitochondria and folate intermediates for nucleotide synthesis. Furthermore, the TCA-cycle intermediates are used as precursors for macromolecule synthesis. For example, the TCA-cycle intermediate citrate is exported to the cytosol for lipid synthesis (Chapter 7). Two-carbon acetyl-CoA and four-carbon oxaloacetate generate the six-carbon citrate. Pyruvate dehydrogenase (PDH) oxidizes pyruvate to generate acetyl-CoA. Pyruvate is produced by glycolysis and/or alanine in the cytosol and, subsequently, is transported into the mitochondria by the pyruvate transporter. Once mitochondrial citrate is generated, it can continue through the TCA cycle and eventually generate oxaloacetate for another round of condensation with acetyl-CoA, resulting in new citrate. However, proliferating cells export a substantial fraction of citrate to the cytoplasm, where it is cleaved by the enzyme ATP citrate lyase (ACLY) to produce acetyl-CoA and oxaloacetate. Acetyl-CoA is used by fatty acid synthase to synthesize lipids. Thus, glucose carbon can eventually become acetyl-CoA in the cytoplasm to produce lipids. The oxaloacetate produced generates malate or aspartate. Malate can be converted into pyruvate by the cytosolic malic enzyme to generate NADPH used for fatty acid synthesis. Aspartate is used for de novo nucleotide synthesis. Aspartate can also be generated within mitochondria from oxaloacetate and transported into the cytoplasm.

In addition to citrate, other TCA-cycle metabolites are used to generate biosynthetic reactions. For example, succinyl-CoA is used for heme synthesis. The export of the TCA-cycle intermediates depletes the cycle of metabolites and the rate-limiting metabolite oxaloacetate. Thus, the cycle must be replenished to generate different TCA-cycle metabolites, resulting in the generation of oxaloacetate and allowing the cycle to continue to function. An important replenishment mechanism is the use of glutamine conversion into glutamate and, subsequently, into α-ketoglutarate (glutaminolysis, see Chapter 4). In vitro and in vivo experiments show that many proliferating cells consume glutamine in significantly greater amounts than other amino acids available to the cell. Glutamine is the most abundant amino acid in plasma. Glutamine is also a required nitrogen donor for the de novo synthesis of both purines and pyrimidines, and in the rate-limiting step catalyzed by glutamine: fructose-6-phosphate amidotransferase, to form glucosamine 6-phosphate, a precursor for *N*- and *O*-linked glycosylation reactions. Finally, glutamine generates glutamate, which is one of the amino acids required for production of the tripeptide GSH, the others being cysteine and glycine.

It is important to note that not all proliferating cells will display glutaminolysis. For example, mouse embryonic stem cells use threonine to feed the TCA cycle. Furthermore, some proliferating cells use pyruvate to generate both acetyl-CoA and oxaloacetate through PDH and pyruvate carboxylase (PC), thus, relieving the necessity to perform glutaminolysis. Acetate can also generate acetyl-CoA for lipid synthesis. There are many inputs into the TCA cycle by amino acids (Chapter 4); thus, it is possible that in vivo there are likely other amino acids, aside from glutamine, that are essential in replenishing the TCA cycle as its metabolites are siphoned off for building macromolecules.

In addition to generating metabolites that build macromolecules, glycolysis and mitochondrial metabolism also produce ATP, which provides the Gibbs free energy (ATP/ADP) to drive unfavorable anabolic reactions. A widely held assumption is that rapidly proliferating cells generate ATP from glycolysis. However, multiple studies in the past decade have carefully evaluated the rate of ATP production from glycolysis and mitochondrial metabolism in proliferating cells and have concluded that although glycolysis does contribute to ATP production, the majority of ATP is derived from mitochondrial metabolism under normal and low-oxygen conditions in most proliferating cells. Endothelial cells are a notable exception and produce significant amounts of ATP from glycolysis. As long as mitochondrial enzymes in the TCA cycle or electron transport complexes are functional, proliferating cells can generate enough ATP through mitochondrial metabolism. Although TCA-cycle intermediates, like citrate, are siphoned off for building macromolecules in the cytoplasm, the constant replenishment of TCA-cycle metabolites by amino acids like glutamine can sustain the TCA cycle to generate the reducing equivalents NADH and $FAHD_2$, which feed into the electron transport chain to produce ATP through oxidative phosphorylation.

## NADPH DRIVES ANABOLISM AND MAINTAINS REDOX BALANCE IN PROLIFERATING CELLS

Many of the energetically unfavorable anabolic reactions in cells are coupled to NADPH/NADP$^+$, and thus proliferating cells have high demand for NADPH production. NADPH is also used to bolster antioxidant capacity (see Chapter 5) in proliferating cells that produce ROS as a by-product of the enhanced oxidative metabolism that occurs in the mitochondria and protein folding in the endoplasmic reticulum. There are multiple sources of NADPH production that proliferating cells can use in the cytosol and mitochondria. The PPP, isocitrate dehydrogenase 1 (IDH1), one-carbon metabolism, and malic enzyme are major cytosolic sources of NADPH. One-carbon metabolism is a major source of mitochondrial NADPH. Different proliferating cells are likely dependent on different cytosolic or mitochondrial sources of NADPH. For example, cancer cells harboring a mutant K-*ras* oncogene use the non-oxidative arm of the PPP to generate ribose-5-phosphate, thus, bypassing the reaction that produces NADPH. In these cells, glutamine-derived malate production is used by cytosolic malic enzyme as the dominant source of NADPH. It is important to note that in any given cell, the dominant NADPH production site will be dependent on the available substrates and enzyme levels of a particular NADPH-generating reaction.

## NUTRIENT UPTAKE THROUGH TRANSPORTERS IS ESSENTIAL FOR PROLIFERATION

An obvious consideration when discussing metabolic requirements of cell proliferation is the uptake of nutrients. Amino acids and sugars are polar molecules and so cannot cross the cell membranes without members of the solute carrier (SLC) family of membrane transport proteins (Fig. 11-3). Glucose and other sugars cross the cell membrane either by facilitated diffusion through glucose transporters (GLUTs) or by active transport through sodium–glucose transporters (SGLTs). Many of the nutrient transporters have common names but do adhere to the specific SLC nomenclature; for example, GLUT1 is SLC2A1. There are 11 SLC families committed to the transport of amino acids. Some of these amino acid transporters important for cell proliferation are linked to the 4F2 heavy chain (4F2hc, CD98, or SLC3A2). 4F2hc is not a nutrient transporter but dimerizes and acts as a chaperone for transporters, such as LAT1 (SLC7A5) and xCT (SLC7A11). The 4F2hc/LAT1 complex exchanges glutamine and other amino acids out of the cell for essential amino acids transported into the cell to stimulate mechanistic target of rapamycin complex 1 (mTORC1), a potent anabolic kinase. The Na$^+$-dependent transporter ASCT2 (SLC1A5) cooperates with 4F2hc/LAT1 by providing glutamine as an export substrate for essential amino acid (EAA) import. xCT exchanges glutamate out of the cell for cystine into the cell, for conversion into intracellular cysteine required for GSH synthesis. SNAT1 (SLC38A1) and SNAT2 transport glutamine into

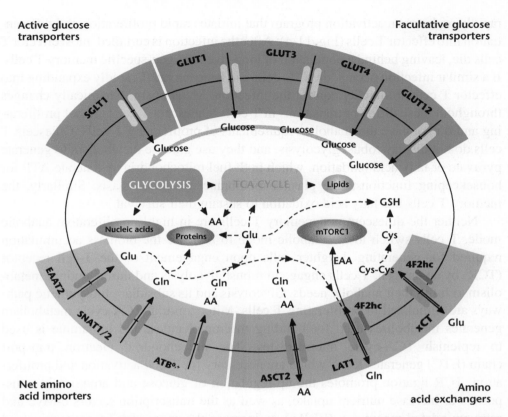

**Figure 11-3.** Cell proliferation requires nutrient transporters. Glucose is imported through SGLTs or GLUTs to fuel glycolysis. Net amino acid transporters, including SNAT1, SNAT2, and ATB⁰,⁺, supply glutamine to fuel the TCA cycle and generation of glutamate for GSH synthesis. Glutamine and other amino acids serve as exchange substrates for transporters, such as ASCT2, 4F2hc/LAT1, and 4F2hc/xCT. LAT1 imports EAA to activate mTORC1. Cystine is transported through xCT to support GSH production. AA, amino acid(s); Cys, cysteine; Cys-Cys, cystine; EAA, essential amino acid (s); Glu, glutamate; Gln, glutamine; GLUT, glucose transporter; GSH, glutathione; SGLT, sodium–glucose transporter. (Adapted from McCracken and Edinger 2013, with permission from Elsevier.)

proliferating cells, whereas EAAT2 (SLC1A2) transports glutamate without exchanging for other amino acids. The expression of sugar and amino acid transporters at the transcriptional and posttranscriptional levels is regulated by growth factor activation of the PI3K (phosphoinositide 3-kinase) signaling pathway and nutrient availability (see below).

## METABOLISM IN T CELLS IS AN EXAMPLE OF NORMAL PROLIFERATING CELLS

T cells respond to antigens; thus, are central regulators of adaptive immune responses. Quiescent naïve T cells, when challenged with an antigen during infection, rapidly undergo an activation program that initiates rapid proliferation and differentiation into effector T cells (Fig. 11-4). After the infection is curtailed, most effector T cells die, leaving behind a population of long-lived antigen-specific memory T cells. If a similar infection occurs, then $T_M$ cells can be reactivated, rapidly expanding into effector T cells to quickly control the infection. Metabolism dynamically changes throughout these different transitions in T cells. Quiescent T cells are not proliferating and do not have the anabolic requirements of proliferating T cells. Quiescent T cells do not display robust glycolysis, and they use glucose metabolism to generate pyruvate or fatty acid oxidation, which both fuel mitochondria to generate ATP for housekeeping functions, like plasma membrane ion homeostasis. Similarly, the memory T cells use fatty acid oxidation to sustain their survival.

Neither the quiescent nor memory T cells are in highly proliferative anabolic mode. T cells switch to an anabolic mode to increase the biomass accumulation required for generating daughter cells upon engagement of the T-cell receptor (TCR) by an antigen. T cells engage in robust glycolysis and mitochondrial metabolism to fulfill their anabolic needs. Glycolysis and its subsidiary biosynthetic pathways are stimulated in proliferating T cells. Mitochondrial TCA-cycle metabolism generates metabolites used for building macromolecules, and glutamine is used to replenish TCA-cycle intermediates. The mitochondrial electron transport chain (ETC) generates ROS, which are necessary for T-cell activation and proliferation. TCR ligation promotes the up-regulation of glucose and amino acid transporters to increase nutrient uptake, as well as the transcription factors c-Myc and estrogen-related receptor α (ERRα), to increase the expression of genes involved

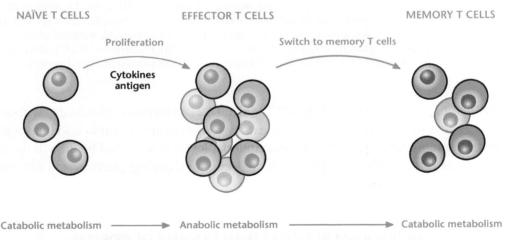

**Figure 11-4.** T cells engage in different types of metabolism depending on their functions. Naïve T cells, following exposure to an antigen during infection, engage in anabolic metabolism, which supports rapid T-effector-cell proliferation and cytokine production. After the infection is curtailed, the effector T-cell response subsides and a few antigen-specific memory T cells remain, which engage in catabolism of nutrients to maintain long-term survival.

in intermediary metabolism. Aside from TCR activation, T cells require CD28 (cluster of differentiation 28) costimulation for T-cell proliferation, which activates PI3K signaling pathways to promote glucose metabolism, as well as multiple anabolic pathways, as discussed in Chapter 10.

Rapidly proliferating CD4$^+$ T cells can differentiate into different effector T-cell lineages, $T_H1$, $T_H2$, and $T_H17$ or a T regulatory lineage (Treg). $T_H17$ cells display a strong glycolytic phenotype caused by hypoxia-inducible factor 1 (HIF-1) activation in these cells, and blocking glycolysis impairs their function. Tregs display a glycolytic and mitochondrial metabolic phenotype. A critical question for the future is to decipher whether metabolism dictates these different T-cell phenotypes or is a consequence of the phenotype.

## ABERRANT ACTIVATION OF SIGNALING PATHWAYS INCREASES ANABOLIC METABOLISM OF PROLIFERATING CANCER CELLS

The metabolism of proliferating cancer cells in part mirrors that of a proliferating T cell, with similar transcription factors, signaling pathways, and nutrients promoting similar anabolic metabolic pathways. The major difference between the two cell types is that cancer cells proliferate in a cell-autonomous manner, whereas T cells are instructed to proliferate by the presence of an antigen. Proliferating cancer cells engage in metabolic pathways that support the massive anabolic program required for the generation of two daughter cells, as discussed above. In mammalian cells, these metabolic pathways are under the control of growth factors and nutrient availability (as discussed in Chapter 10). Cancer cells aberrantly drive these signaling pathways that control metabolism. In particular, the gain of oncogenes and loss of tumor suppressor genes, two key features of cancer cells, co-opt metabolism into an anabolic program. During the evolution of a tumor, cancer cells have the metabolic plasticity to adjust to the different microenvironments they encounter, ranging from abundant to limiting nutrients.

Cancer is a heterogeneous collection of diseases with genomic heterogeneities between histologically similar tumors. Thus, it is of no surprise that cancer cells display metabolic heterogeneity and there is no universal cancer metabolism model sufficient to describe the metabolic changes required to support tumor growth. However, what is consistent among a spectrum of tumors is that to grow they need (1) sufficient energy (ATP and NADPH), (2) building blocks to generate macromolecules, and (3) redox balance maintenance due to their high-production ROS. Cancer cells can use diverse pathways to harness these three important constituents to support growth. For example, some tumors might acquire fatty acids in large amounts from the extracellular milieu and use it to generate membranes and fuel mitochondrial metabolism through fatty acid oxidation. In contrast, certain tumors will use glucose and glutamine to generate de novo lipid synthesis. Although glucose and glutamine have been linked as major fuel sources for cancer cells, it is likely that a range of other amino acids are also important sources of carbon and nitrogen molecules required for building macromolecules, as well as generating sufficient energy for growth. Mitochondria and glycolysis are two major sources of ATP in cancer cells, and, depending on their microenvironment nutrient availability and genetic lesions, these cells can engage in either one or both of these pathways for ATP generation. As discussed previously, there are multiple sources of NADPH and cancer cells can call on any or all of these sources.

Finally, cancer cells use NADPH oxidases and the mitochondrial ETC to produce the ROS, which maintain many of the signaling pathways in an activated state. Cancer cells also show high protein folding levels in the endoplasmic reticulum, and this process also generates ROS as a by-product. The high rate of ROS production is counterbalanced by an equally high rate of antioxidant activity in cancer cells to maintain redox balance (Fig. 11-5). Cancer cells show different rates of ROS production and induce a multitude of antioxidant proteins and so are heterogeneous in their antioxidant profile and capacity. If cancer cells do not control their ROS levels, then they are susceptible to oxidative stress-induced cell death. The signaling pathways responsive to ROS are proximal to the locations of ROS generation, allowing activation of these pathways despite the high overall antioxidant activity in cancer cells that protects against oxidative stress-induced cell death.

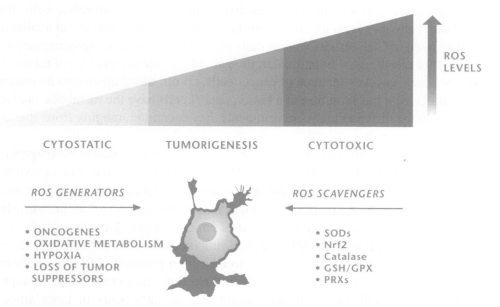

**Figure 11-5.** Cancer cells maintain redox balance. Cancer cells have an elevated production of ROS, which activates proximal signaling pathways necessary for tumorigenesis. The high rate of ROS production is counterbalanced by an equally high rate of antioxidant activity in cancer cells to maintain redox balance. GPX, glutathione peroxidases; PRXs, peroxiredoxins.

Despite the genetic and metabolic heterogeneity of tumor cells, it is worth mentioning a few recurring pathways that co-opt metabolism to support the growth of tumors (Fig. 11-6). In normal cells, growth factors, through engagement of their receptors, activate PI3K and its downstream pathways AKT and mTOR, which promote a robust anabolic program (see Chapter 10). Tumor cells have gain-of-function mutations in PI3K or loss-of-function mutations in PTEN, the negative regulator of PI3K, that alleviate the necessity of growth factor–dependent

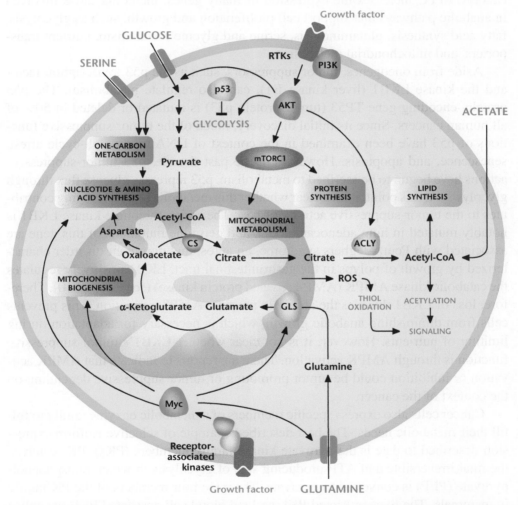

**Figure 11-6.** Signaling pathways that regulate cancer cell metabolism. Tumor cells have gain-of-function mutations in PI3K or loss-of-function mutations in PTEN, the negative regulator of PI3K, that alleviate the necessity of growth factor–dependent signaling. MYC, which is aberrantly activated in a variety of cancers, increases the expression of many genes involved in anabolic pathways that support cell proliferation and growth. Metabolism can also regulate signaling, in part, through production of mitochondrial ROS and acetyl-CoA. GLS, glutaminase; ACLY, ATP-citrate lyase; CS, citrate synthetase. (Adapted from Ward and Thompson 2012.)

signaling. In fact, there are multiple oncogenes and tumor suppressors identified in the PI3K signaling pathway network, and aberrant activation of this pathway is among the most frequent alterations seen in a diverse set of cancers. For example, the oncogenic K-*ras*, which is frequently found in lung, colon, and pancreatic cancers, uses the PI3K pathway to stimulate anabolic metabolism for tumor growth. Oncogenic K-*ras* also uses the proto-oncogene MYC to promote an anabolic program to support growth. MYC is also aberrantly activated by chromosomal translocations, gene amplification, and single-nucleotide polymorphisms in a variety of cancers. MYC increases the expression of many genes, including those involved in anabolic pathways that support cell proliferation and growth, such as glycolysis, fatty acid synthesis, glutaminolysis, serine and glycine metabolism, nutrient transporters, and mitochondrial metabolism.

Aside from oncogenes, tumor suppressors, such as the p53 transcription factor and the kinase LKB1 (liver kinase B1), can also regulate metabolism. The p53 protein–encoding gene TP53 (tumor protein p53) is mutated or deleted in 50% of all human cancers. Since its initial discovery, much of the tumor-suppressive functions of p53 have been examined in the context of DNA repair, cell-cycle arrest, senescence, and apoptosis. However, in the past decade, p53 tumor-suppressive actions have begun to be ascribed to metabolism. p53 represses glucose flux through glycolysis. At this point, it is unclear whether this metabolic reprogramming contributes to the tumor-suppressive activity of p53. The serine-threonine kinase LKB1 is notably mutated in lung adenocarcinoma, and germline mutations of this gene are associated with Peutz–Jeghers syndrome, an autosomal dominant disorder characterized by growth of polyps in the gastrointestinal tract. LKB1 positively regulates the catabolic kinase AMPK (AMP-activated protein kinase) (see Chapter 10). Therefore, loss of LKB1 disables the ability to promote AMPK activation. This prevents cells from diminishing anabolic growth, which is necessary for adaptation during limiting of nutrients. However, it is not clear whether LKB1's tumor-suppressive function is through AMPK activation. There are reports to indicate that AMPK activation or inhibition could be tumor promoting or tumor suppressive depending on the context of the cancer.

Cancer cells also express specific members of a metabolic enzyme family to fulfill their metabolic needs. The best-described example of selective isoform expression described to date is the pyruvate kinase family members (PKs). PKs catalyze the final irreversible and ATP-producing step of glycolysis in which phosphoenolpyruvate (PEP) is converted to pyruvate. There are four members of the PK family in mammals. The liver-restricted PKL and red blood cell–restricted PKR are splice variant isoforms encoded by the PK-LR gene. PKM1 and PKM2 are splice variant isoforms encoded by the PK-M gene. The M1 and M2 isoforms differ by a single exon and share ~96% sequence identity at the amino acid level. Most proliferating cells, including cancer cells, express the M2 isoform of the enzyme rather than its M1 splice variant. Cancer cells engineered to express PKM1 instead of PKM2 display reduced tumor-forming ability, underscoring the importance of PKM2 for cancer progression. PKM2 alternates between a dimer that shows low catalytic activity and a highly active tetramer. Paradoxically, however, PKM2 has an enzymatic activity half that of PKM1 and is typically found inactive in vivo. This is due, in part, to tyrosine phosphorylation specific to the M2 isoform, a modification that inhibits its activity by disrupting tetramer assembly (Fig. 11-7). Furthermore, the

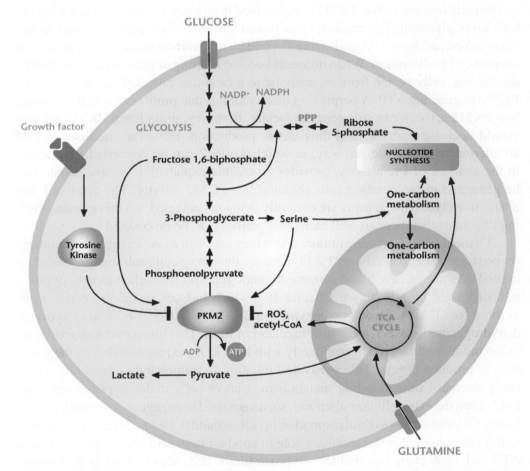

**Figure 11-7.** Proliferating cancer cells express PKM2 to regulate glycolysis. PKM2 alternates between a dimer that shows low catalytic activity and a highly active tetramer. Proliferating cells typically have reduced PKM2 activity, in part, because of tyrosine phosphorylation, a modification that inhibits its activity by disrupting tetramer assembly. Acetylation at specific lysine residue and ROS-induced oxidation at a cysteine residue within PKM2 also reduce its activity. When nutrients are not limiting and growth factor–dependent signaling pathways are active, PKM2 is maintained in an inactive state, thus allowing the buildup of glycolytic intermediates that funnel into subsidiary pathways, such as the pentose phosphate pathway (PPP) and serine-dependent one-carbon metabolism pathway, to support cell proliferation. When nutrients become limiting or growth factor signaling diminishes, cells activate PKM2 to generate ATP. (Adapted from Ward and Thompson 2012.)

increased availability of cytosolic acetyl-CoA in proliferating cells causes acetylation of PKM2 at lysine 305 to further reduce its activity. ROS also can oxidize PKM2 at specific cysteine residues, resulting in its inactivation. In contrast, PKM1 is constitutively active and not regulated by ROS or tyrosine kinase signaling. One model to explain why PKM2 is advantageous to cancer cells is that PKM2 fluctuates between an inactive and active state to regulate metabolism, depending on nutrient and growth factor availability. When nutrients are not limiting and growth factor–dependent signaling pathways are active, PKM2 is maintained in an inactive state, thus allowing the buildup of glycolytic intermediates that funnel into subsidiary pathways, such as the hexosamine pathway, PPP, and serine-dependent one-carbon metabolism pathway to support cell proliferation. When nutrients become limiting or growth factor signaling diminishes, cells switch from an anabolic to a catabolic program and also activate PKM2 to generate ATP. A perplexing observation is that proliferating cells showing low PKM2 activity can still generate lactate. In theory, diminished PKM2 activity should decrease pyruvate and thus lactate production. However, the discovery of an alternative glycolytic pathway, in which the PEP can be converted into lactate in the absence of PK activity, provides a possible explanation of how lactate can be generated in proliferating cells showing low PKM2 activity. The details of this alternative glycolytic pathway are currently being investigated, yet this exciting finding illustrates that there are still metabolic pathways to be discovered.

A salient feature of many tumors is that they reside in a low-oxygen environment (hypoxia) ranging from 0% to 2% $O_2$ because the tumor cell proliferation rate often exceeds the rate of new blood formation (angiogenesis). The adaptation to hypoxia is coordinated by HIF-1 (see Chapter 10), which induces metabolic genes, such as those for GLUTs, glycolytic enzymes, and LDHA. HIF-1 also induces pyruvate dehydrogenase kinase-1, a negative regulator of PDH. This limits pyruvate oxidation in the mitochondria and concomitantly with high LDH expression diverts the pyruvate to produce lactate. This pyruvate limitation to mitochondria does not necessarily decrease mitochondrial metabolism. Cancer cells under hypoxia scavenge lipids from the extracellular milieu and, so, do not need to engage in de novo lipid synthesis. Oxygen begins to limit respiration in cells around 0.3% $O_2$. Therefore, hypoxic tumor cells above this threshold are able to conduct fatty acid oxidation to generate ATP and TCA-cycle metabolites. Hypoxic tumor cells also engage in glutamine-dependent reductive carboxylation to generate TCA-cycle metabolites (discussed in the next section). Finally, there are tumors that display constitutive activation of HIF1 and HIF2 under normoxic conditions through a variety of mechanisms, including (1) hyperactivation of mTOR, (2) loss of von Hippel–Lindau protein (pVHL), (3) accumulation of ROS, and (4) accumulation of the TCA-cycle metabolites succinate or fumarate because of cancer-specific mutations in succinate dehydrogenase (SDH) or fumarate hydratase, respectively (discussed in the next section).

The combination of hypoxia, gain of oncogenes, loss of tumor suppressors, and high rate of protein folding in the endoplasmic reticulum results in high levels of ROS production. As noted above, cancer cells increase their antioxidant proteins to maintain redox balance. The major mechanism by which cancer cells increase their antioxidant proteins is through activating the transcription factor nuclear factor erythroid 2-related factor 2 (NRF2). Normally, NRF2 interacts with Kelch-like ECH-associated protein 1 (KEAP1), thus targeting NRF2 for proteasomal degradation. Elevated ROS oxidizes redox-sensitive cysteine residues on KEAP1, resulting in dissociation of KEAP1 from NRF2. Subsequently, NRF2 translocates to the nucleus, heterodimerizes with the small protein MAF, and binds to antioxidant-responsive elements within the regulatory regions of multiple antioxidant genes. Aside from elevated ROS, signaling pathways, such as the ERK1/2 mitogen-activated protein kinase pathway and PI3K, can activate NRF2. Furthermore, certain tumor cells have mutations of KEAP1 that result in constitutive activation of NRF2 and its target antioxidant genes. The loss of NRF2 in cancer cells increases oxidative stress to levels that trigger cell death, resulting in diminished tumorigenesis. This observation has led to the idea that increasing oxidative stress selectively in cancer cells might be a viable therapeutic strategy.

## GENETIC ALTERATIONS IN SPECIFIC METABOLIC ENZYMES CAN DRIVE TUMORIGENESIS

In recent years, it is increasingly appreciated that metabolic enzymes act genetically as tumor suppressors or oncogenes. Initial recognition of genetic alterations in metabolic enzymes driving cancer was the identification of loss-of-function germline heterozygous mutations in SDH and fumarate hydratase (FH) at the turn of this century. Loss of heterozygosity of SDH occurs in certain cases of paraganglioma, pheochromocytoma, and FH in leiomyoma and certain cases of renal cell carcinoma. The loss of FH and SDH prevents the TCA cycle from functioning properly and, thus, the cells rely on glycolysis for ATP generation. These cells are the exception, not the rule, for tumor cells relying exclusively on glycolysis for ATP generation. Although the canonical TCA cycle is not functioning in these cells, SDH- and FH-deficient tumors are able to use glutamine to generate α-ketoglutarate; a reverse TCA-cycle reaction ensues in which α-ketoglutarate is converted by isocitrate dehydrogenase 2 (IDH2) to generate isocitrate and, eventually, citrate. Subsequently, citrate can be exported into the cytosol to generate acetyl-CoA and oxaloacetate for de novo lipid and nucleotide synthesis, respectively. This process is glutamine-dependent reductive carboxylation, in which a carbon molecule from $CO_2$ is used as a substrate (not a product) to generate citrate. Cells showing an electron transport deficiency also show reductive carboxylation. SDH and fumarate hydratase tumors generate high levels of succinate and fumarate, respectively, which can inhibit 2-α-ketoglutarate-dependent dioxygenase enzymes (Fig. 11-8). These include prolyl hydroxylases (PHDs), TET (ten-eleven translocation) DNA hydroxylases, and Jumonji-domain

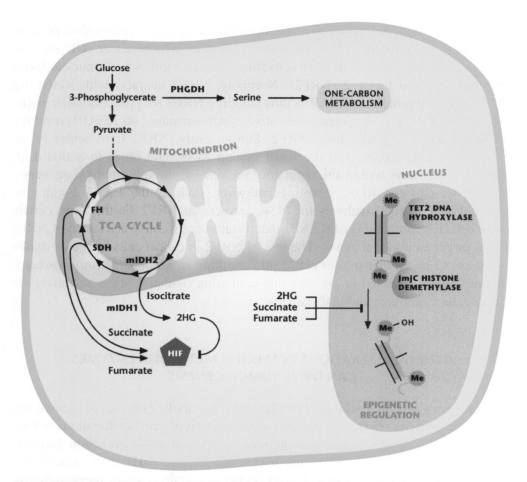

**Figure 11-8.** Alterations in certain metabolic enzymes drive cancer. Mutations in TCA-cycle enzyme succinate dehydrogenase (SDH) and fumarate hydratase (FH) occurs in certain cancers, resulting in accumulation of succinate and fumarate, respectively. Mutations in cytosolic IDH1 or mitochondrial IDH2 occur in certain cancers and cause accumulation of 2-hydroxyglutarate (2HG). Succinate, fumarate, and 2HG inhibit histone demethylases and TET DNA hydoxylases. Succinate and fumarate can activate HIF, whereas R-2HG inhibits HIF. Elevated expression of wild-type phosphoglycerate dehydrogenase (PHGDH) in certain cancers increases serine biosynthesis to fuel one-carbon metabolism necessary for NADPH production and folate intermediates necessary for nucleotide synthesis. (Adapted from Cantor and Sabatini 2012, by permission from the American Association for Cancer Research.)

(JmjC) histone demethylase enzymes. The inhibition of these enzymes increases HIFs, as well as hypermethylation of DNA and histones. SDH and FH tumor cells also produce high levels of ROS that promote activation of HIFs and cell proliferation. Fumarate binds directly to GSH, thus increasing ROS levels. However, the ROS levels are maintained in a range that is compatible with survival and proliferation by fumarate binding to Keap1, the negative regulator of NRF2. This maintains redox balance in FH-deficient cells. FH and SDH tumors show the metabolic plasticity of tumor cells and how metabolites can have functions beyond their canonical roles in intermediary metabolism.

In recent years, whole-genome sequencing has led to the identification of cytosolic IDH1 and mitochondrial IDH2 mutations in a fraction of acute myeloid leukemias, gliomas, and chondrosarcomas. To date, there have been mutations found in mitochondrial isocitrate dehydrogenase 3 (IDH3) linked to cancer. All three enzymes catalyze the oxidative decarboxylation of isocitrate to produce $CO_2$ and α-ketoglutarate. IDH1 and IDH2 generate NADPH, whereas IDH3 produces NADH, which is used for ATP generation. IDH1 and IDH2 are homodimers and IDH3 is a heterotrimer.

IDH1 and -2 mutations are somatically acquired where one allele remains wild type (WT) and the other allele mutated (MUT) at a single catalytic arginine. The WT/MUT IDH1 or IDH2 dimer allows for normal fast production of α-ketoglutarate and NADPH, therefore not perturbing metabolism of these mutant isocitrate dehydrogenase (IDH) cancer cells. But the WT/MUT dimer, at a slower rate, uses NADPH and converts α-ketoglutrate to the (R)-enantiomer of 2-hydroxyglutarate (R-2HG), eventually allowing R-2HG to accumulate to high levels. R-2HG is barely detectable under normal conditions. R-2HG inhibits α-ketoglutarate dioxygenases, TET2 DNA hydroxylases, and JmjC demethylases, which are central regulators of epigenetics. However, unlike fumarate and succinate, it does not inhibit PHDs, but rather stimulates, resulting in repression of HIF activity. The emerging model is that IDH mutations through R-2HG mediate their tumorigenic effects, in part, through epigenetic dysregulation. There are likely to be other mechanisms beyond epigenetics that fully explain R-2HG-dependent tumorigenesis. Currently, 2HG is being used as a biomarker for disease monitoring, and inhibitors specific to IDH1/2 are beginning to undergo clinical trials. Interestingly, R-2HG production through unknown mechanisms is also observed in certain cases of breast tumors that do not necessarily show IDH mutations, and also in tumor cell lines exposed to hypoxia in vitro.

Aside from mutations in metabolic enzymes, certain cancers show elevated expression of metabolic enzymes. To date, the best examples are the elevated expression of phosphoglycerate dehydrogenase (PHGDH) and glycine decarboxylase (GLDC). PHGDH catalyzes the conversion of 3-phosphoglycerate to 3-phosphohydroxypyruvate in the first step of the serine biosynthesis pathway (Fig. 11-8). The PHGDH protein is elevated in certain cases of malignant breast cancer and melanomas, resulting in increased flux of glucose carbon through the serine biosynthesis pathway that branches from glycolysis. Serine fuels the one-carbon metabolism necessary for NADPH production (see Chapter 5, Fig. 5-6) and folate intermediates necessary for nucleotide synthesis (see Chapter 9, Fig. 9-4) and methylation reactions (see Chapter 8, Fig. 8-11). Suppression of PHGDH in tumor cells that show high levels of PHGDH results in a decrease in serine synthesis and cell growth. GLDC overexpression was identified as a molecular signature of tumor-initiating cells (TICs) of non–small cell lung cancer (NSCLC), but not bulk NSCLC cells, again highlighting the issue of intratumor metabolic heterogeneity. GLDC is a component of the glycine cleavage system, which catalyzes glycine degradation to produce folate intermediate (see Chapter 5, Fig. 5-6). GLDC suppression diminishes proliferation and tumorigenicity of NSCLC TICs. Enhanced GLDC expression increases in pyrimidine biosynthesis and makes these cells susceptible to low doses of methotrexate, an inhibitor of the folate cycle (see Chapter 9, Fig. 9-11), highlighting how metabolic profiling might dictate therapy.

## METABOLISM IS BEING TARGETED FOR CANCER THERAPY

During the past decade, there has been excitement about targeting metabolism as a rational strategy for the treatment of cancer. However, it is important to recognize that this approach is not new, and antimetabolite drugs targeting nucleotide synthesis that were developed in the mid-20th century were the first widely successful class of drugs. They include analogs of pyrimidines (e.g., 5-fluorouracil), purines (e.g., azathioprine), and antifolates (methotrexate). Some of these drugs are still used in the treatment of leukemia, lung, breast, and colorectal cancers, and others, like methotrexate and azathioprine, are also used for treatment of inflammatory conditions, such as rheumatoid arthritis. However, these drugs also have adverse effects, primarily because they do not distinguish between highly proliferating normal cells and cancer cells, which require de novo nucleotide synthesis. Thus, it is not surprising that these antimetabolite drugs suppress the immune system and affect tissues that actively turn over.

Current efforts are deciphering metabolic enzymes on which cancer cells display a higher dependency than normal proliferating cells do. This is a daunting task because the metabolism of normal proliferating cells and cancer cells display many similarities in the metabolic and signaling pathways that they use. Unless these new cancer metabolic therapies can distinguish between malignant and nonmalignant cells, then the same types of toxicity that plague conventional antimetabolites could complicate the therapeutic targeting of cancer metabolism. Another consideration is that cancer cells display metabolic plasticity. Cancer cells can develop resistance to inhibition of a particular metabolic pathway through expression of alternate isoforms, up-regulation of alternate pathways, or the use of adjacent cells, such as adipocytes, to provide precursors for the biosynthesis of macromolecules. Finally, the metabolic heterogeneity observed among tumors of the same subtype, or even within a single tumor, can further make it challenging to target specific cancer metabolism.

So, what are possible metabolic enzymes that are good candidates for cancer therapy? The obvious candidate is targeting mutant IDH1 or IDH2 enzymes in tumors showing IDH1 or IDH2 mutations, respectively. Currently, there are drugs that can distinguish between mutant IDH1 and IDH2 versus their WT counterparts. However, these are the exception, rather than the rule, because very few tumors display gain-of-function mutations in metabolic enzymes. Current research efforts are designed to find metabolic enzymes that are overexpressed in certain cancer cells compared with normal cells. Much effort has been devoted to targeting glucose metabolism. One example is the overexpression of hexokinase (HK) II in many tumors. HKs catalyze the first step of glycolysis and, hence, are a potentially attractive target. There are four mammalian HKs (HKI–IV). HKI is the ubiquitously expressed isoform, whereas HKII is expressed in insulin-sensitive tissues, such as muscle and adipose. Preclinical studies show that HKII inhibition could be an effective cancer therapy. Likewise, targeting enzymes overexpressed in the glutaminolysis pathway are appealing for cancer therapy. Preclinical studies show that targeting glutaminase, the first step in glutaminolysis, could be effective against certain cancers. It is likely that targeting both glucose and glutamine metabolism would be effective compared with targeting either pathway because of metabolic plasticity. For example, overexpressing PC, thus allowing glucose-derived pyruvate to feed the TCA cycle by generating oxaloacetate and acetyl-CoA, could compensate for glutaminolysis inhibition. Glycolysis inhibition could be compensated by glutaminolysis generation of TCA-cycle metabolites that serve as precursors for gluconeogenesis. Importantly, it remains to be determined whether preventing glucose and/or glutamine metabolism is effective therapy for certain cancers without incurring toxicity similar to the classical antimetabolites targeting nucleotides synthesis.

The inhibition of glucose or glutamine metabolism in certain cancer cells significantly decreases NADPH and GSH levels, both decreasing antioxidant capacity, resulting in dramatic increases in ROS levels to induce cell death. Cancer cells generate increased ROS as by-products of their increased metabolism. ROS have a well-defined role in promoting and maintaining tumorigenicity. Yet, most clinical trials have failed to show beneficial effects of administering antioxidants in a variety of cancer types; in certain trials, antioxidants have been shown to promote cancer. This effect might be caused by the fact that many therapeutic antioxidants are not effective in targeting the low levels of mitochondrial ROS generated proximally to signaling pathways, which are required for tumorigenesis.

More recent studies have focused on disabling antioxidants selectively in cancer cells, thus, raising their ROS levels to thresholds that induce death. As mentioned earlier, cancer cells overexpress an array of antioxidant proteins, in part, through the activation of NRF2 to maintain ROS levels that allow protumorigenic signaling pathways to be activated without inducing cell death. In fact, studies have shown that disabling antioxidant mechanisms triggers ROS-mediated cell death in a variety of cancer cell types. This reliance on antioxidants may represent the cancer cell's "Achilles heel," as nontransformed cells produce less ROS and, therefore, are less dependent on their detoxification. It is important to note that loss of NRF2 diminishes

## BOX 11-1. METFORMIN: AN ANTIDIABETIC DRUG REPURPOSED AS AN ANTICANCER AGENT

One agent that, recently, has been the focus of many cancer metabolism therapeutic studies is the repurposing of the antidiabetic drug metformin as an anticancer agent. Metformin is widely used to treat patients with type 2 diabetes mellitus. Metformin suppresses liver gluconeogenesis, thereby reducing glucose release from the liver. In several recent retrospective studies, investigators have observed an association between metformin use and diminished tumor progression in patients suffering from different types of cancers (Box 11-1, Fig. 1). These data have prompted more than 100 ongoing prospective clinical trials to determine the efficacy of metformin as an anticancer agent. However, the underlying mechanism by which metformin diminishes tumor growth are just beginning to be unraveled. The two not mutually exclusive mechanisms by which metformin reduces tumor growth are (1) at the organismal level, in which it reduces the levels of circulating insulin, a known mitogen for cancer cells, and (2) in a cell-autonomous manner by targeting mitochondrial electron transport at complex I. The organismal mechanism is based on the observation that some cancer cells express insulin receptors, which are potent stimulators of PI3K pathways. Thus, metformin's inhibitory effect on hepatic gluconeogenesis to reduce circulating insulin levels would decrease tumor

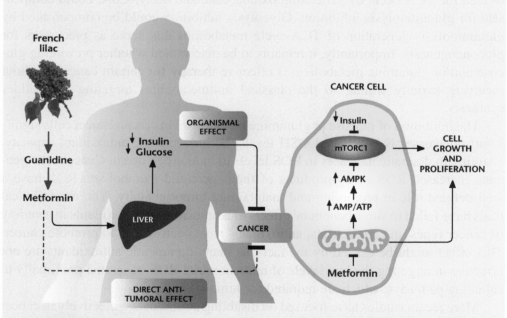

**Box 11-1, Figure 1.** Metformin reduces tumorigenesis through multiple mechanisms. The biguanide metformin, modeled after guanidine derivatives, was first isolated from the French lilac *Galega officinalis*. Currently, metformin is the most commonly used drug worldwide to treat patients with type 2 diabetes mellitus. Metformin suppresses liver gluconeogenesis, thereby reducing glucose release from the liver. Recently, metformin has been repurposed as an anticancer agent. There are two not mutually exclusive mechanisms by which metformin reduces tumor growth: (1) at the organismal level, in which it reduces the levels of circulating insulin, a known mitogen for cancer cells, and (2) in a cell-autonomous manner, by targeting mitochondrial electron transport at complex I. (Adapted from Birsoy et al. 2012, by permission from Macmillan Publishers Ltd.)

growth by diminishing insulin receptor activation of the PI3K pathway. The cell-autonomous mechanism is dependent on whether cancer cells express organic cation transporters (OCTs) to promote metformin into cells, resulting in complex I inhibition. Metformin has a safe toxicity profile in normal tissues because it accumulates only in a few normal tissues that express OCT transporters, such as liver. In contrast, traditional complex I inhibitors, such as rotenone, readily accumulate in any normal or cancer cell and, consequently, are highly toxic. Metformin inhibition of mitochondrial complex I within cancer cells is likely to diminish tumor growth through multiple mechanisms, including preventing mitochondrial ROS generation required for mitogenic signaling, increasing AMP levels to activate the catabolic kinase AMPK, and inducing cell death when tumor cells become limiting for glucose and cannot generate ATP through glycolysis. Thus, the combination therapy of metformin with clinically used PI3K inhibitors that reduce glucose uptake and glycolysis is likely to be more efficacious than metformin alone. Cancer cells have a wide range in expression of OCTs and insulin receptors. The ongoing clinical trials using metformin as an anticancer agent should assess the expression levels of OCTs and insulin receptors in the tumors to identify those likely to be susceptible to metformin, which are those with the highest expression.

multiple antioxidant defense systems and, therefore, makes multiple types of ROS increase at a threshold that invokes damage to cancer cells. However, loss of a specific antioxidant defense system might result in elevation in ROS levels to levels below the threshold that causes damage. In this scenario, the elevated ROS levels hyperactivate signaling pathways to promote tumorigenesis, as observed during the loss of peroxiredoxin I, which increases tumorigenesis. Nevertheless, diminishing antioxidant capacity in cancer cells is likely to synergize with traditional chemotherapeutic agents, also known to increase ROS levels. It is perplexing that many people with cancer continue to take large concentrations of antioxidants in conjunction with chemotherapy, thus negating the potential benefits of their chemotherapy.

In conclusion, metabolism has regained its importance in cancer biology after being ignored for decades. The observations that signaling pathways, oncogenes, and tumor suppressors regulate metabolic enzymes has reenergized cancer metabolism. A pleasant consequence of reemergence of cancer metabolism is that it has emphasized the importance of metabolism in other fields, including inflammation, stem cells, and cell biology. The recent advances in metabolomics (see the Appendix) have revealed new metabolic pathways and allowed a more comprehensive assessment of tumor metabolism in patients. Furthermore, the discovery of certain mutations in metabolic enzymes linked to cancer has bolstered the idea that changes in metabolism are not simply a consequence of proliferating cells, but that they play a causal role in tumorigenesis. The current excitement surrounding cancer metabolism research is that the metabolic enzymes critical for cancer cell proliferation and survival are potential targets for cancer therapy. A big challenge, going forward, will be to decipher which metabolic enzymes necessary for tumorigenesis have a favorable therapeutic index.

## REFERENCES

Birsoy K, Sabatini DM, Possemato R. 2012. Untuning the tumor metabolic machine: Targeting cancer metabolism: A bedside lesson. *Nat Med* **18:** 1022–1023.

Cantor JR, Sabatini DM. 2012. Cancer cell metabolism: One hallmark, many faces. *Cancer Discov* **2:** 881–898.

Deberardinis RJ, Sayed N, Ditsworth D, Thompson CB. 2008. Brick by brick: Metabolism and tumor cell growth. *Curr Opin Genet Dev* **18:** 54–61.

McCracken AN, Edinger AL. 2013. Nutrient transporters: The Achilles' heel of anabolism. *Trends Endocrinol Metab* **24:** 200–208.

Vander Heiden M. 2011. Targeting cancer metabolism: A therapeutic window opens. *Nat Rev Drug Discovery* **10:** 671–684.

Ward PS, Thompson CB. 2012. Metabolic reprogramming: A cancer hallmark even Warburg did not anticipate. *Cancer Cell* **21:** 297–308.


## ADDITIONAL READING

Chandel NS, Tuveson DA. 2014. The promise and perils of antioxidants for cancer patients. *N Engl J Med* **371:** 177–178.

Dang CV. 2012. Links between metabolism and cancer. *Genes Dev* **26:** 877–890.

DeBerardinis RJ, Thompson CB. 2012. Cellular metabolism and disease: What do metabolic outliers teach us? *Cell* **148:** 1132–1144.

Fan J, Kamphorst JJ, Mathew R, Chung MK, White E, Shlomi T, Rabinowitz JD. 2013. Glutamine-driven oxidative phosphorylation is a major ATP source in transformed mammalian cells in both normoxia and hypoxia. *Mol Syst Biol* **9:** 712.

Gorrini C, Harris IS, Mak TW. 2013. Modulation of oxidative stress as an anticancer strategy. *Nat Rev Drug Discov* **12:** 931–947.

Hardie DG, Alessi DR. 2013. LKB1 and AMPK and the cancer-metabolism link—Ten years after. *BMC Biol* **11:** 36.

Israelsen WJ, Dayton TL, Davidson SM, Fiske BP, Hosios AM, Bellinger G, Li J, Yu Y, Sasaki M, Horner JW, et al. 2013. PKM2 isoform-specific deletion reveals a differential requirement for pyruvate kinase in tumor cells. *Cell* **155:** 397–409.

Keith B, Johnson RS, Simon MC. 2011. HIF1α and HIF2α: Sibling rivalry in hypoxic tumour growth and progression. *Nat Rev Cancer* **12:** 9–22.

Klempner SJ, Myers AP, Cantley LC. 2013. What a tangled web we weave: Emerging resistance mechanisms to inhibition of the phosphoinositide 3-kinase pathway. *Cancer Discov* **3:** 1345–1354.

Losman JA, Kaelin WG Jr. 2013. What a difference a hydroxyl makes: Mutant IDH, (R)-2-hydroxyglutarate, and cancer. *Genes Dev* **27:** 836–852.

Patra KC, Wang Q, Bhaskar PT, Miller L, Wang Z, Wheaton W, Chandel N, Laakso M, Muller WJ, Allen EL, et al. 2013. Hexokinase 2 is required for tumor initiation and maintenance and its systemic deletion is therapeutic in mouse models of cancer. *Cancer Cell* **24:** 213–228.

Pearce EL, Pearce EJ. 2013. Metabolic pathways in immune cell activation and quiescence. *Immunity* **38:** 633–643.

Pollak M. 2013. Potential applications for biguanides in oncology. *J Clin Invest* **123:** 3693–3700.

Yang M, Soga T, Pollard PJ. 2013. Oncometabolites: Linking altered metabolism with cancer. *J Clin Invest* **123:** 3652–3658.

# 12

# Future Pathways of Metabolism Research

## INTEGRATION OF METABOLISM WITH DIET AND ENVIRONMENT

**Navdeep S. Chandel**
*Northwestern Medical School, Chicago, Illinois*

T HE MAIN OBJECTIVE OF THIS BOOK IS TO highlight the integration of cellular metabolism to cellular signaling. As I look at the past 25 years of conducting research on cellular metabolism, the most enthralling aspect of research has been trying to understand the cross talk between metabolism and signaling and how it impacts biological outcomes, such as proliferation, differentiation, and metabolic adaptation. In this final chapter, I have asked my contemporary scientific colleagues and friends to reflect on what they are excited about in metabolism studies. A consistent theme among these essays is the excitement in trying to understand the integration of metabolism with biology, physiology, and pathology.

My own interest, going forward, is to continue to understand the incorporation of metabolism with the rapidly expanding, interesting, new aspects of cell biology. Recently, a particular area that has fascinated me is the connection between metabolism and the increase in certain diseases. This is particularly important because, in the past 25 years, there has been an increase in autoimmunity, allergies, diabetes, obesity, and chronic obstructive pulmonary disease. The curse of cancer is rising alarmingly as more countries modernize. Furthermore, as the aging population increases, this is accompanied by an increase in neurodegenerative diseases. Geneticists have been working diligently to find the genes that are causing this upsurge in different pathologies. However, to date, genetic variants do not explain the increase in many of these modern diseases. Maybe we will find the gene that has caused the sudden escalation in peanut allergies. However, I think there is an alternative explanation.

I would argue that most of our genes have been optimally selected for the diet, environment, and behavior prevalent through human evolution. What has changed

211

is not our genes but our diet (high-fructose consumption), environment (increase in pollution), and behavior (smoking and lack of physical activity). As a result, our genes are in conflict with the brave new world in which we currently exist. A simple solution would be to modify diet, environment, and behavior. Yet, this proves to be socially, economically, and politically challenging. So this leaves us to figure out how modifications in diet, environment, and behavior elicit the devastating consequences manifested for a range of diseases. Metabolism is poised to initially sense these modifications in diet, environment, and behavior and drive changes in gene expression, which ultimately cause phenotypic changes that manifest as pathologies. Thus, metabolism is the driver and genes are the passengers in causing modern diseases, and understanding metabolism could provide therapeutic targets to alleviate the increase in diseases.

## METABOLISM AND AGING

**Anne Brunet**
*Stanford University, Palo Alto, California*

The past 20 years have established a clear role for conserved metabolic pathways, the insulin-FOXO, AMP-activated protein kinase (AMPK), and target of rapamycin (TOR) pathways and mitochondrial activity in the regulation of aging in species ranging from yeast to mammals. The field has also started to identify the tissues in which these metabolic pathways primarily act to regulate aging and how different tissues can communicate to one another. An exciting question for the future is whether changes in metabolic status could have long-lasting influences on the organism. For example, could nutrient status changes early in life induce changes later in life and, perhaps, even in next generations if these changes affect the germline? Understanding how metabolic switches are integrated and induce long-term changes in the cell circuitry or changes to stable chromatin states should produce new insights into this question. In the future, it will also be exciting to uncover how cells with different strategies—differentiated cells, somatic regenerative stem cells, "immortal" germline cells—respond to changes in nutrient status and what molecules are used to communicate between cells and tissues in a non-cell-autonomous manner. It will likewise be important to understand the "non-organismal-autonomous" influence on metabolism. For example, how does the microbiome, which is known to be important for nutrient absorption, interact with cellular mechanisms of the host to regulate organismal metabolism and aging? In worms and flies, males and females have recently been found to communicate via pheromones in a way that affects both metabolism and life span. It will be interesting to further understand the impact of one sex on the other for the regulation of metabolism and life span and determine whether this phenomenon is conserved in higher species. Overall, it will be exciting to uncover how the metabolic regulation of aging interfaces with the other external forces known to regulate organismal life span.

## CARBON SUBSTRATE CHOICE IN THE PROGRAMMING OF CELL FATE AND FUNCTION

**Nika N. Danial**
*Harvard Medical School, Boston, Massachusetts*

Different cell states have distinct anabolic and catabolic needs that can be fulfilled by processing specific carbon substrates. Recent advances in metabolomics and metabolic flux analysis, combined with high-throughput interrogation of the genome and proteome, have provided some insights into molecular determinants of cellular fuel choice and its regulation of cellular fate and function. Emerging evidence indicates that the cells' fuel choice can influence transitions between quiescence and proliferation, produce resistance or sensitivity to oxidative stress, facilitate DNA and tissue repair, and allow metabolic adaptations to nutrient changes. Furthermore, metabolism of particular carbon substrates can influence cell identity and behavior through programmatic alterations in gene expression and epigenetic modifications. Understanding how these fuel switches are controlled and defining their specific metabolic outputs will provide a molecular handle on modulating cell behavior in normal physiology and in pathologic conditions. In particular, it would be important to determine whether the reprogramming of cellular fuel choice is a cause or a consequence of transitions in and out of quiescence, lineage specification, and cellular adaptation to stress. Within the context of fuel choice, substrate supply and substrate access are likely important control points that may be heavily influenced by metabolic compartmentalization/channeling, spatial organization of metabolic enzymes in macromolecular complexes, and mitochondrial architecture. Our molecular understanding of these connections is currently limited, providing exciting opportunities for future discoveries. Furthermore, it is likely that, beyond genetically encoded metabolic programs, additional mechanisms can cue cells to preferentially metabolize a particular carbon substrate when they have access to multiple fuels. The mechanisms underlying this "fuel competition" will provide important molecular insights into cellular fuel choice.

## FUTURE PERSPECTIVES OF CANCER METABOLISM

**Eyal Gottlieb**
*Beatson Institute, Glasgow, Scotland, United Kingdom*

For many years, the metabolic changes in cancer were studied alongside the genetic and molecular signaling associated with the disease, but these changes were considered peripheral to the cancer research field. Over the past decade, we have witnessed a surge in experimental technological and pharmacological approaches that have signaled the emergence of cancer metabolism research as a major discipline. This surge was driven mainly by a twofold recognition: first, that many oncogenic events directly control metabolism and, second, that several metabolic enzymes are oncogenes and tumor suppressors. We are still at the early stages of research in the field of cancer metabolism. Nevertheless, this field shows undeniable promise for understanding tumorigenesis and the development and design of new cancer therapies.

Cancer metabolism is one of the most encouraging areas emerging in medical research. An explosion in technological advances, such as high-throughput metabolomics and metabolic computational models, allows us to study the metabolic adaptations and rewiring that contribute to cancer cells' growth and survival. Although the use of metabolic changes for diagnostic positron emission tomography scanning exemplifies the impact that this area can have on clinical practice, the full extent of the clinical applications of this field is only now beginning to come to light. Cancer metabolomics today can be compared with the significant advances in cancer management resulting from cancer genomics studies in the last decades. There is great potential here for the rapid identification of new metabolic therapeutic targets and specific cancer cell vulnerabilities and for the development of treatment methods. This field also offers the opportunity for finding the best approach for using existing metabolic drugs as cancer treatments and for applying metabolomics technologies to improve prognosis and diagnosis.

## FOLLOW THE ELECTRONS!

**Michael P. Murphy**
*Medical Research Council Mitochondrial Biology Unit, Cambridge, United Kingdom*

During the 1970s Watergate era in the United States, the way to the truth was to "follow the money." Similarly, one way to make progress in metabolism is to "follow the electrons." The last few years have shown that metabolism is central to cell function and fate in unexpected ways. Where previously we focused on identifying genes and gene products, now we look to changes in metabolites and fluxes. In parallel, ROS modulate metabolism by redox signaling, and protein posttranslational modifications, such as acetylation and thiol oxidation, respond to redox couples: NADH/NAD, GSH/GSSG, and acetylCoA/CoA. However, in making sense of metabolic changes, we often force what we find into the rigid pathways we learned in "kindergarten" and consider changes in redox signaling, redox couples, and the associated posttranslational modifications as separate processes. Of course, they are all intimately interlinked and interdependent.

How best, then, to consider this unity and frame hypotheses that generate informative experiments? I think it is useful to see metabolism as a flow of electrons from carbohydrate and fat, passing through multiple channels, a bit like a river delta. A main distributary is glycolysis and the mitochondrial respiratory chain to reduce oxygen and generate ATP. Other major channels funnel electrons through biosynthetic pathways, such as the pentose phosphate pathway and meandering braids of the

TCA cycle, to create building blocks for the cell and to store electrons as NADPH/NADP for biosynthesis. As the electrons flow through these well-worn channels, the fluctuating levels in redox couples, such as NADH/NAD, acetylCoA/CoA, or ubiquinol/ubiquinone, continually influence the rest of the metabolic network by feedback signals, posttranslational modifications, and also the spillover of electrons to generate ROS. In aging, cancer, or diabetes, these channels are disrupted and the interconnections and flows are out of balance, explaining why metabolic and ROS disruption appears so often in these disorders. So, to make sense of the complexity of metabolism and how it is altered in pathology, I suggest that we "follow the electrons" as they flow through the cell.

## METABOLIC LINKS TO IMMUNITY AND INFLAMMATION

**Jeffery Rathmell**
*Duke University, Durham, North Carolina*

Metabolic pathways fuel and support every aspect of physiology, and a major transition in metabolic research has been the recognition that classic metabolic tissues are not alone in their regulation by cell metabolism. The view had been that most cells were able to acquire the nutrients they needed based on consumption, release of feedback inhibitions, and then nutrient uptake. This process is now clearly understood to be far more complex, as each tissue and cell type has specific metabolic needs and regulatory mechanisms to ensure these demands are met. Understanding these cell- and stimuli-specific mechanisms for controlling metabolic pathways and how those pathways, in turn, influence other cell functions are enormous areas for the future of metabolic research.

One emerging field based on these questions is immunometabolism. Metabolic influence over the immune system has been appreciated for a more than a century, as malnutrition often led to increased infection and disease, whereas obesity promotes inflammation and is associated with autoimmunity. The immune system has a wide variety of cells with specific functions and demands. Normally, pathogen encounter leads to a series of regulatory events and metabolic reprogramming as cells are activated to provide protection. How precisely metabolic reprogramming is regulated and what it provides to immune cells, however, is a wide-open question. Importantly, metabolic reprogramming means different things to different cells, and the most inflammatory and immunologically suppressive cells often use entirely opposite programs. These distinctions mean that inflammatory and suppressive cells have different levels of specific enzymes and metabolites that can contribute directly to immunological function through biosynthesis, energy, and even nonmetabolic signaling functions. The links between immune metabolic reprogramming and functions are poorly understood, but have enormous implication and potential to comprehend and potentially treat autoimmunity, cancer, and infectious disease.

## MASTER AND SERVANT: THE RECIPROCAL CONTROL
## OF METABOLISM AND CELLULAR DECISIONS

Jared Rutter

*University of Utah School of Medicine, Salt Lake City, Utah*

When I grew up as a biologist, it was certainly my impression that protein-based regulatory systems made all the important cellular decisions. Kinases, G proteins, and transcription factors all talk to each other and, voilà!, cell behavior and fate are decided. Of course, I believed that metabolism was important, but only as an obedient servant. When the "brains" of the cell make a decision, the metabolic network must be marshaled to ensure that the necessary building blocks or energetic resources are available. It is increasingly clear to me, however, that the metabolic state of the cell has a direct input into the decision-making apparatus of the cell and plays an active role in most decisions. For example, manipulations designed to specifically alter the metabolic program have major effects on cell fate during development and oncogenesis. Stem and cancer cells tend to show commonalities in their metabolic profile, which is distinct from the programs used by most nonproliferating, differentiated cells. These metabolic schemes are optimized to the behavior of the cell. A proliferative cell has a higher demand for carbon-based synthetic intermediates, like nucleotides, amino acids, and lipids, than does a nonproliferative cell. As a result, it is intuitive that a cell, having made the decision to be proliferative, would adapt its metabolism to meet the demands. Exciting and unexpected recent results have convinced me, however, that metabolism is driving the decision at least as much as the converse. There are many, many specific examples that exemplify this principle and certainly others will be discovered. In the end, I believe that understanding how metabolic inputs control cellular decisions might provide us with our best opportunities for manipulating those decisions therapeutically.

## NUTRIENT SENSING

David M. Sabatini

*Whitehead Institute, Massachusetts Institute of Technology, Cambridge, Massachusetts*

Animals have specialized tissues that sense nutrient levels and convey them via hormones to all the cells of the organism. Such hormonal systems, like the glucose-induced release of insulin from the pancreas, allow animals to coordinate the responses of different organs to periods of fasting or feeding. Increasingly, we appreciate that, in addition to these non-cell-autonomous mechanisms, most cells have cell-autonomous means for sensing nutrient and energy levels. Several pathways, such as those anchored by the mechanistic target of rapamycin complex 1 (mTORC1), AMPK, and GCN2 kinases, respond to changes in amino acid, glucose, and energy levels to control the balance between anabolic and catabolic processes, like protein synthesis and autophagy. AMPK and GCN2 are not only the central nodes of their respective pathways, but also the sensors, with AMPK detecting the AMP/ATP ratio and GCN2 detecting the uncharged transfer RNAs that accumulate during amino acid starvation. In contrast, mTORC1 does not appear to directly sense nutrient or energy sufficiency, but instead is downstream of a complex signaling system that detects an array of environmental cues, including amino acids, glucose, energy, and insulin.

How mTORC1 senses and integrates these diverse inputs has become a topic of great interest, and recent work has started to decipher the mechanisms involved. The Rag and Rheb GTPases have essential, but different, roles in mTORC1 pathway activation, with the Rags dictating the subcellular localization of mTORC1 and Rheb turning on its kinase activity. Nutrients, particularly amino acids, activate the Rag GTPases, which then recruit mTORC1 to the lysosomal surface where Rheb resides. Upon insulin withdrawal, the tuberous sclerosis complex tumor suppressor translocates there and inhibits mTORC1 by inactivating Rheb. Thus, the Rag and Rheb inputs converge at the lysosome, forming two halves of a coincidence detector that ensures that mTORC1 activation occurs only when both are active. A lysosome-associated complex containing the multisubunit Ragulator and vacuolar ATPase is upstream of the Rag GTPases and key for mTORC1 activation by nutrients. However, how these complexes are regulated is not understood, and the identity of the nutrient sensor(s) is unknown. This basic system has been worked out mostly in human cells in culture, and how it is adapted in vivo to allow tissues to respond to different types or levels of nutrients remains to be determined. Information on how amino acids are sensed in particular tissues may have clinical utility. For example, in elderly people, the sensing of amino acids by mTORC1 in skeletal muscle is defective, and therapies that can reverse this may be useful for promoting muscle anabolism and, thus, slowing aging-associated loss of muscle mass.

## MITOCHONDRIAL METABOLIC SIGNALS CONTROL NUCLEAR GENE EXPRESSION

**Gerald S. Shadel**
*Yale School of Medicine, New Haven, Connecticut*

Specialized cell types and tissues are defined at one level by unique metabolic profiles that enable distinct functional capabilities. They also possess the need and ability to respond to physiological and environmental cues in a stereotypical manner. It is usually assumed that a key underlying driver of this type of specificity is differential nuclear gene expression based on unique transcriptional and/or epigenetic signatures. To tailor metabolism and metabolic responses in this way, and to avoid inappropriate responses that could lead to cell dysfunction or cell death, requires real-time readouts of metabolic and stress signals. Mitochondria are perfectly positioned to provide this type of information, and recent advances show that signals emanating from these essential organelles orchestrate adaptive changes in nuclear gene expression.

Mitochondria are central hubs of intermediary metabolism, calcium homeostasis, and ROS production. They are also platforms for stress signaling pathways, including those that respond to hypoxia, energy deprivation, and pathogen infection. A salient example is mitochondrial ROS signaling pathways that promote longevity in budding yeast and worm model systems. In yeast, this involves mitochondrial ROS signaling through the DNA damage-sensing kinase ATM (Tel1p in yeast) and inactivation of the jumonji histone demethylase Rph1p specifically at subtelomeric chromatin. In worms, mitochondrial ROS stress is transmitted through the apoptotic signaling cascade to increase stress resistance. Interestingly, in both cases, the mitochondrial ROS signal is co-opting these stress-signaling pathways in noncanonical ways (i.e., not by activating a DNA damage or proapoptotic response). These examples illuminate a new paradigm that may be developing—metabolic signaling operates through previously identified stress pathways but uses them in unique ways to engender specific nuclear gene expression outcomes.

One can imagine, then, in addition to ROS, many other mitochondrial and metabolic signals exist that relay transcriptional and epigenetic changes in the nucleus. For example, iron, TCA-cycle intermediates (e.g., succinate and fumarate), and redox cofactors, like $NAD^+$, are candidate molecules for which there is already strong evidence for signaling roles. An important future area in metabolic research is to delineate and understand the full complement of these mitochondrial metabolic signaling pathways, what the key molecular messengers and signal transducers are, and how they are involved in human disease and aging. For these pathways to hold potential as therapeutic targets, we need to understand how mitochondria and associated stress-signaling pathways operate differentially in specific cell types and tissues and during key developmental windows.

## REBIRTH OF CANCER METABOLISM: DECADE TWO

Reuben J. Shaw
*Salk Institute, La Jolla, California*

We are currently in the exhilarating stage of decoding the precise molecular links and rate-limiting signaling events that gate the metabolic pathways of the cell. In mammalian cells, most metabolic enzymes are not just a single enzymatic activity shown on a classic metabolic pathway chart, but, rather, are often encoded by multiple genes within a gene family, most bearing multiple alternative splice forms that tune enzymatic activity up and down. Which isoforms of the various metabolic enzymes are most important for a given tumor type and which phosphorylation and acetylation events act as dominant inhibitory or activating marks on those enzyme isoforms are only now being discovered. In response to distinct hormonal and nutrient cues, kinases and deacetylases act to coordinate the activity of intersecting metabolic pathways through rapid fine-tuning of the activity of specific steps of five to10 different metabolic processes. Overlaying the metabolite control of metabolic pathways essential to proliferation with the posttranslational modifications of the enzymes involved is an ongoing area of investigation. Synthesizing that with larger cell biological processes, including the coordinating of metabolism via mitochondrial fission, fusion, mitophagy, and apoptotic decisions, will be another area for development.

The emergence of several bona fide metabolic enzymes (isocitrate dehydrogenase 1, -2 [IDH1, IDH2], FH, succinate dehydrogenase) as tumor suppressors and oncogenes, coupled with the discoveries that a number of oncogenes and tumor suppressors count metabolic adaptation as one of their main physiological functions (myc, von Hippel–Lindau tumor suppressor, mTOR, liver kinase B1/STK11), means that in 10 years, it will not be accurate to have a metabolic wiring chart of the cell that does not have genes frequently mutated in cancer on it. Conversely, no future chart of human cancer genes will omit the regulation of cellular metabolism and energetics as one of the core processes coordinated with the cell cycle and growth and survival pathways. A current goal, then, becomes to further refine our understanding of the rate-limiting metabolic enzymes involved in survival and proliferation of tumors with different genetic driver mutations and from different tissues of origin, which retain a lot of their often unique metabolic controls from the tissue-specific context in which the tumor arose. Being able to use mouse genetic models to perform genetic and pharmacological validation of these metabolic vulnerabilities will be part of the next phase, as cancer metabolism once again moves from the biochemistry benches to the oncology clinic.

## TOWARD A MODERN UNDERSTANDING OF CELL METABOLISM

**Matthew G. Vander Heiden**
*Massachusetts Institute of Technology, Cambridge, Massachusetts*

Luminaries working in the first half of the last century described much of what is known about cell metabolism. Their fundamental discoveries included how cells couple the oxidation of carbon to capture free energy as ATP and how hydrolysis of high-energy phosphate bonds can drive otherwise unfavorable biochemical reactions to support cell homeostasis. These foundational studies also described the biochemical pathways used to generate the amino acids, lipids, nucleic acids, and carbohydrates that cells use to store energy and make up all living organisms. Most of these details were uncovered before the revolution in molecular biology that allowed genetic manipulation of cells and organisms, and long before genome sequencing provided an enzyme parts list for cells. As a result, our understanding of cell metabolism is incomplete, and the coming decades offer tremendous opportunities to apply modern tools to answer basic questions about how cell metabolism is regulated in cells, tissues, and organisms.

Most scientists learn metabolism at the stage of their career when biology is presented as factual knowledge. Subsequently, many view the machinery providing energy and materials for complex biological phenomena, like signaling, gene expression regulation, growth, division, and movement, as background functions. However, work over the last decade has shown that metabolic regulation is intimately linked with countless aspects of cell biology. As a result, the study of metabolism has reentered the mainstream of scientific investigation, and the coming years promise to provide a wealth of new information about how metabolism interfaces with normal and disease physiology. These studies will also inform us how cell metabolism can be harnessed to increase food production, create biofuels, and improve medical science.

Some key questions about cell metabolism that will be answered in the coming years include a better understanding of how metabolic regulation is tuned to match the physiological state and enable different cell functions. Textbook descriptions of metabolism are derived mostly from studies of differentiated tissues and stationary-phase microorganisms, in part because these systems have allowed ready access to the quantities of material needed for biochemical analysis using classic approaches. Modern work to understand cancer metabolism has illustrated that central carbon metabolism is regulated differently to support a proliferative state, and work to understand how these findings apply across tissues in the context of whole-organism physiology will undoubtedly reveal new insights. Another area of active interest is how outputs of cell metabolism, other than ATP, are controlled. For instance, cellular redox balance is poorly understood yet differs across cell states, is tightly controlled, and has a profound impact on cell function. Finally, advances in systems biology will enable an understanding of metabolic networks in ever-increasing detail. Past work to describe metabolism as a system have focused on single-compartment systems, such as red blood cells or prokaryotes. The inability of these descriptions to predict many biological outcomes illustrates our incomplete understanding of metabolic networks. New approaches to resolve compartment-specific processes and model complex systems will lead to more discoveries in this field. The study of cell metabolism is entering a second golden age and promises to provide answers to some of the most fundamental questions in all of biology.

## METABOLISM AS A THERAPEUTIC TARGET IN CANCER

**Kate Yen**
*Agios Pharmaceuticals, Cambridge, Massachusetts*

The discovery, in 1924, by Otto Warburg that tumor cells undergo aerobic glycolysis to meet their energy requirements was the first evidence that cancer cells have altered cellular metabolism. This rewiring functions to meet the metabolic requirements of tumor cells, and many studies have now shown that genetic alterations can have both a direct and indirect effect on cellular metabolism. Somatic hot-spot mutations in IDH enzymes, IDH1 and IDH2, were first identified through a genome-wide mutational analysis in glioblastoma. Subsequently, these mutations have been found in multiple tumor types, including acute myeloid leukemia (AML), cholangiocarcinoma, and chondrosarcoma. These mutations impair the ability of IDH1and IDH2 to catalyze the conversion of isocitrate to α-ketoglutarate (αKG), and confer a novel gain-of-function catalytic activity leading to the reduction of αKG to the metabolite (*R*)-2-hydroxyglutarate (2HG). High levels of 2HG impair αKG-dependent dioxygenases, including both histone and DNA demethylases, leading to epigenetic rewiring of the cell, a block in cellular differentiation, and tumorigenesis. Recently, inhibition of this neomorphic activity has been shown to reverse this block in cellular differentiation, as well as lead to a reduction in tumor growth in preclinical studies. Finally, clinical development of AG-221, a selective small-molecule inhibitor of mutant IDH2, has shown early clinical proof of principle in patients with AML and myelodysplastic syndrome, highlighting that inhibition of altered metabolic pathways in tumor cells may offer a promising new avenue for therapeutic intervention. Although these early studies targeting mutant IDH are promising, our understanding of how cancer metabolism impacts tumor growth is still in its infancy. The paucity of hot-spot mutations in other metabolic enzymes, as well as the complexity of data acquisition and analysis, has contributed to the challenge in identifying other metabolic targets in cancer. Going forward, linking cancer genetics to a dependence on metabolic pathways will hopefully help identify, a priori, patients most likely to respond to targeted metabolic intervention and, hopefully, improve their clinical outcome.

# APPENDIX

# Analyzing Metabolism in Biological Systems

Ralph J. DeBerardinis[1]

*Children's Medical Center Research Institute, University of Texas,*
*Southwestern Medical Center*

THERE ARE A VARIETY OF METHODS FOR measuring metabolites and the activity of metabolic pathways in biological systems. Some of these techniques are efficient enough to screen entire human populations for levels of metabolites that have particular relevance to disease. Others can be used to evaluate metabolism in vivo noninvasively, without the need to obtain tissue or body fluids. Several of these methods are introduced below.

A key point is that there are several dimensions of metabolic activity that should be considered in the experimental design. In this sense, analysis of metabolism is analogous to the ways one might study traffic patterns. One conceptually simple way to begin to analyze traffic is to ask how many cars are present at a given time, what types of cars, and in what proportions to each other. For example, one could consider making a detailed list of every car in a parking lot and studying how that list changes over time. Similarly, one could also carefully measure the steady-state abundances of a large number of metabolites and ask how these levels change among various biological settings (e.g., normal vs. diseased tissue). This type of quantitative steady-state analysis is commonly called metabolomics.

However, measurements of steady-state abundance cannot provide a complete picture of dynamic systems. Cars move along roads and highways, and metabolites progress along biochemical pathways. In both cases, some assessment of "movement" is necessary to understand the system. It is difficult to know, based on the abundance of a metabolite like phosphoenolpyruvate, whether the tissue is undergoing glycolysis or gluconeogenesis. A second dimension of metabolic analysis, therefore, involves identifying which pathways are active in a given biological

---

[1] Ralph.Deberardinis@UTSouthwestern.edu

system. This is most commonly accomplished by presenting the system with a precursor containing an isotopic label that can be transmitted to downstream molecules as a function of metabolic activity. At the end of the labeling period, metabolites are extracted and analyzed for the abundance and location of the label, and this information is used to infer pathway activity. This type of analysis is commonly called isotope tracing.

A third and related form of analysis is to determine the rate of movement. Even if the overall pattern of traffic has been established, the rates of these pathways can vary considerably from moment to moment—think of how differently traffic flows on the same highway between rush hour and midnight. Flux rates of metabolic pathways may vary by more than an order of magnitude, depending on the biological state, so that even identifying the entire set of active pathways would fail to produce a comprehensive view of metabolism. Metabolic flux analysis (MFA) seeks to describe, either quantitatively or on a ratiometric basis, the rates of metabolic pathways throughout a network. This process usually involves a combination of isotope tracing and computational modeling of the subsequent labeling data, ideally incorporating both the isotopic enrichment of each target and the position of each label within target molecules.

## TWO TECHNOLOGIES ARE WIDELY USED TO STUDY METABOLISM

A detailed discussion of the technologies available to analyze metabolism is well beyond the scope of this Appendix. However, it is important to understand the strengths and limitations of the two most widely used analytical approaches: nuclear magnetic resonance spectroscopy (NMR) and mass spectrometry (MS). Both techniques are applied routinely to study all three dimensions of metabolism discussed above. Both approaches have the capacity to quantify many metabolites, and both can be used to trace the fates of various isotopes relevant to intermediary metabolism, particularly $^2H$, $^{13}C$, and $^{15}N$. The techniques differ from each other in several important ways. NMR has two major advantages over MS. First, it has the potential to detect metabolites and enzyme-catalyzed metabolic activities noninvasively in vivo. This is not an option for MS, which requires the user to obtain tissue or body fluids for analysis. Second, NMR provides positional specificity of the isotope tracer within target molecules. In some cases, knowing precisely where one or more isotopes are positioned within a metabolite is required to differentiate between two metabolic pathways. Although it is possible to obtain similar information via MS, this process is cumbersome and often impractical compared with NMR. On the other hand, MS has several important advantages over NMR. First, the sensitivity of MS is generally much greater than that of NMR, both for metabolite detection and analyzing isotope enrichment. The outstanding sensitivity of high-quality, modern mass spectrometers make it feasible to detect hundreds to thousands of individual

metabolites from small biological samples. From a variety of distinct MS technologies, one may choose an instrument optimized for sensitivity or mass accuracy,
or a combination of the two, and these instruments may be coupled to various chro-
matography systems to achieve outstanding separation of metabolites from complex
biological samples. Examples of both NMR and MS applications in metabolic analysis are discussed below, with some of the advantages of each method highlighted.

## MEASURING METABOLITE ABUNDANCE

Planning a metabolomics experiment requires first defining the scope of the metabolite pools to be monitored. These experiments can broadly be classified as "targeted"
or "untargeted/global." In targeted metabolomics, a set of preselected metabolites is
analyzed with high specificity and sensitivity. The goal is usually to identify those
metabolites, ideally from a common pathway, that can differentiate between two or
more biological states (e.g., normal vs. disease). MS, particularly tandem MS, which
is well suited to maximize sensitivity, is the most common analytical platform for targeted metabolomics. Anywhere from a few dozen to several hundred metabolites may
be monitored in a single experiment. An example of a successful application of targeted metabolomics is a study by Thomas Wang, Robert Gerszten, and colleagues
that led to the identification of plasma metabolites associated with the risk of type 2
diabetes (Wang et al. 2011). By profiling amino acids and other polar metabolites
from more than 2000 individuals profiled for 12 years, these investigators determined
that modest elevations of aromatic and branched-chain amino acids conferred a significant risk of developing diabetes over the period of study.

The high specificity and throughput of targeted metabolomics has made it possible to use this approach to screen individuals rapidly for altered levels of metabolites with direct relevance to health. In fact, millions of newborn babies every year are
subjected to MS-based targeted analysis of a few dozen plasma metabolites in newborn screening programs. The goal of these programs is to identify babies with
inborn errors of metabolism before the onset of symptoms so that treatment can be
initiated as early as possible. The first example of a metabolic disease subjected
to screening across large populations was phenylketonuria (PKU), a disorder involving defective catabolism of the amino acid phenylalanine. Abnormal accumulation
of phenylalanine by-products impairs brain development, and over time untreated
children with PKU develop severe intellectual disabilities. However, detecting the
disease within the first few weeks of life—before the onset of symptoms—allows
a carefully tailored diet reduced in phenylalanine to be introduced to the babies.
With lifelong dietary modification, PKU patients can achieve normal brain development. Since the 1990s, the application of MS has made it possible to screen for many
disorders concurrently using just a few drops of blood obtained about 24 hours after
birth. Many states in the United States now quantify a few dozen metabolites in

virtually every newborn baby, facilitating the diagnosis of some 30 individual diseases, all of which have a treatment whose timely institution correlates with improved long-term health.

In global MS-based profiling, the investigator does not need to define a set of target metabolites before the experiment. Rather, the goal is to obtain information about as many metabolites as possible, even if a significant fraction of those detected are not assigned to known metabolic pathways. A successful example of global profiling is the work that led to the discovery of an "oncometabolite," or a metabolite associated with cancer. A high fraction of tumors of the brain (gliomas) contain mutations in the genes encoding either of two isoforms of the metabolic enzyme isocitrate dehydrogenase (IDH1 or IDH2). In either case, the mutations are confined to the tumor and present as only one allele of the gene, leading to a situation in which the cancer cell expresses one normal (wild-type) and one mutant copy. A metabolic perturbation of some kind was presumed to result from these common mutations, although it was unclear initially how the mutations altered enzyme function. In 2009, unbiased metabolomic profiling was used to identify differences between cancer cells engineered to express either mutant or wild-type IDH1 protein (Dang et al. 2009). This analysis showed that cells with mutant IDH1 accumulated massive amounts of the (*R*) enantiomer of 2-hydroxyglutarate (*R*-2HG), a metabolite normally present at very low levels in tumors. This was a highly compelling finding because patients with a rare genetic condition predisposing them to high levels of the (*S*) enantiomer of this same metabolite were known to have an elevated risk of brain tumors. This metabolomic finding has stimulated a tremendous amount of investigation into the transforming properties of *R*-2HG, and the development of drugs designed specifically to inhibit mutant forms of IDH1 and IDH2.

As an extreme but clinically powerful example of targeted metabolite detection, magnetic resonance spectroscopy (MRS), has been used to measure *R*-2HG noninvasively in human brain cancer. The millimolar levels of *R*-2HG in brain tumors with mutant isoforms of IDH1 or IDH2 make it possible to quantify this molecule by detecting its protons using MRS. This technique is similar to magnetic resonance imaging and requires neither sampling of the tissue nor injection of metabolic probes. Spectroscopic detection of *R*-2HG correlates virtually 100% with the presence of an IDH1 or IDH2 mutation (Andronesi et al. 2012; Choi et al. 2012; Pope et al. 2012). It may be possible to use this technique to monitor *R*-2HG levels during cancer therapy.

## TRACING METABOLIC PATHWAYS

Isotopes have long been used to study metabolic processes in living systems. Initially, radioactive isotopes (radioisotopes), which could be detected by their emission of energy, were used. György de Hevesy first used radioisotopes of lead and bismuth in plant and animal studies, leading to his being awarded the Nobel Prize in

Chemistry in 1943. Much of what we know about metabolism is derived from studies using radioisotopes, and radioisotopes are still commonly used in the laboratory. Tiny amounts of radioisotopes are also used as tracers in metabolic imaging experiments in human patients, where they may be detected using positron emission tomography. However, if a large amount of label is needed, or if the experiment calls for the analysis of multiple pathways simultaneously, radioisotope studies have largely been replaced by stable isotopes. Unlike radioisotopes, stable isotopes do not undergo spontaneous decay; this eliminates the safety concerns that limit application of radioactivity in animal and human studies.

Either NMR or MS can be used to detect the transfer of stable isotopes through metabolic systems. Isotopes with an odd number of protons and/or neutrons contain an intrinsic nonzero spin that enables them to be detected by NMR. For example, the dominant naturally occurring form of carbon atoms, $^{12}C$, does not contain a nonzero spin and is, therefore, not visible by NMR. The next most common form of carbon is the isotope $^{13}C$, which contains one extra neutron and is visible by NMR. Approximately 1.1% of all carbon is in the form of $^{13}C$, meaning that even a pure "unlabeled" standard would be visible by NMR spectroscopy by virtue of the naturally occurring $^{13}C$. Standards enriched for $^{13}C$, many of which are commercially available, are nearly 100 times easier to detect by NMR. Following this logic, introducing a nutrient (e.g., glucose) enriched for $^{13}C$ in one or more positions allows the fate of the labeled carbon to be tracked throughout a metabolic network using NMR.

Using MS to track isotope transfers through metabolic pathways simply exploits that fact that mass spectrometers can easily distinguish between two forms of a metabolite differing by their isotopic composition. For example, naturally occurring glucose has a nominal mass of 180. Approximately 1.1% of each of glucose's six carbons will be $^{13}C$ instead of $^{12}C$, meaning that a predictable fraction of the total pool will have a mass of 181. A pure standard of glucose, in which one of glucose's carbons is replaced with $^{13}C$, would have a predominant mass of 181, and a fraction of the pool will have a mass of 182 because of naturally occurring $^{13}C$ at other positions. It is essential to account for the "natural" abundance of isotopes when analyzing metabolism by either NMR or MS.

Using stable isotopes to trace metabolism is conceptually straightforward (Fig. A-1). The precursor pool (e.g., glucose) is partially or completely labeled with a stable isotope. A common experimental setup in cultured cells would be to replace unlabeled glucose in tissue culture medium with glucose labeled in one or more positions with $^{13}C$, and then feed a culture of cells with this labeled medium. At the end of the culture period, metabolites are extracted from the cells and analyzed for $^{13}C$ enrichment. A wealth of information can be derived from these experiments, including the fraction of each metabolite containing the label, the occurrence of multiple labels in the same metabolite, and the position of labels within the metabolite. All of these parameters encode information about the metabolic pathways active during the culture period.

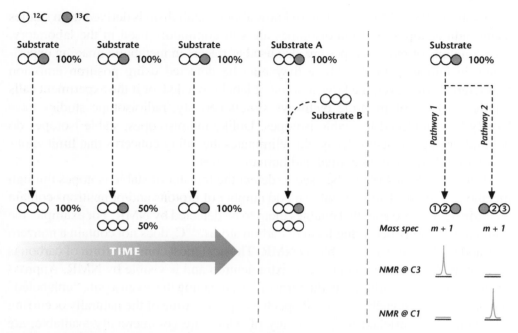

**Figure A-1.** Principles of isotope tracing. Isotope tracing experiments monitor the replacement of atoms within a product pool with a label initiating on a substrate. The example illustrated here tracks the flow of a carbon from a three-carbon substrate into the pool of a three-carbon product. "Unlabeled" carbons (white circles) have an atomic mass of 12 ($^{12}$C) and the "labeled" carbons (gray circles) have an atomic mass of 13 ($^{13}$C). At the instant the experiment begins, 100% of the substrate pool is labeled and none of the product pool is labeled. Over time, $^{13}$C is progressively transferred into the product pool such that $^{12}$C is gradually replaced by $^{13}$C. In a very simple system, like the one depicted on the *left*, in which the labeled substrate is the only source of carbon for the product, and the entire product pool turns over, eventually 100% of the product pool becomes labeled with $^{13}$C. In the more complex system illustrated in the *middle*, a second substrate (Substrate B) can also provide carbon to the product pool. In this case, the product pool never becomes 100% labeled, even after infinite time. The fraction of the product pool containing $^{13}$C is determined by the relative contribution of the two substrates. An even more complex system is illustrated on the *right*. Here, the labeled substrate enters two distinct pathways that can deliver $^{13}$C to the product pool, but each pathway delivers the label to a different position in the product (carbon 3 [C3] in pathway 1 and carbon 1 [C1] in pathway 2). MS detects the presence of one additional mass unit ($m + 1$), but unless additional information is incorporated, it is insensitive to the presence of two distinct pathways. NMR, in contrast, directly differentiates between the two positions of $^{13}$C and would, therefore, easily alert the user that two pathways are operating simultaneously.

Isotope tracing is a highly versatile technique that allows multiple different tracers to be applied in parallel or combination to the same biological system. An example of the use of multiple isotope tracers, analyzed by both NMR and MS, to understand metabolic preferences and compensatory pathways in cancer cell lines is in Cheng et al. (2011). Although $^{13}$C is the most widely used isotope for analysis of central pathways of intermediary metabolism, other isotopes, particularly $^{2}$H and

$^{15}$N, are also used and can be highly informative. Tracing using $^2$H has recently produced novel insights into how cancer cells produce NADPH for redox homeostasis and reductive biosynthesis (Fan et al. 2014).

Similar logic applies to experiments in live subjects, animal or human. In one study by Teresa Fan, Andrew Lane, and colleagues, human patients with lung cancer were given a bolus of [U-$^{13}$C]glucose before surgical resection of their tumors (Fan et al. 2009). Metabolites extracted from the tumor and surrounding normal lung were analyzed for $^{13}$C enrichment by NMR and MS. This analysis produced the first view of intermediary metabolism in a human tumor growing in its native microenvironment. Although the rates of individual pathways were not quantified, isotope tracing revealed that the tumors contained enhanced levels of many $^{13}$C-labeled metabolites relative to the normal lung tissue. These metabolites were from several distinct metabolic pathways, including glycolysis, the Krebs cycle, and anaplerosis. Altogether, the data suggested that all of these pathways were more active in tumors than in normal lung cells.

## USING ISOTOPE TRACERS TO QUANTIFY METABOLIC FLUX

A complete description of metabolism must contain quantitative information about the rate of pathways within the network. Individual pathways within the same cell differ dramatically from each other. Therefore, although it is useful to know whether two pathways are active under a particular set of conditions, one would clearly also want to know which pathways were highly active and which were barely moving. Rates of individual pathways also differ according to the biological state, and activation or suppression of metabolic pathways subserves essential biological functions. For example, T cells display low rates of glycolysis in the naïve state and massively increase their glycolytic rate on stimulation with mitogenic signals. Enhanced glycolysis enables these cells to produce the energy and macromolecules required for growth and proliferation. Thus, although glycolysis is active in both the naïve and stimulated states, altered flux through the pathway is one of the most prominent metabolic effects of mitogenic stimulation.

MFA refers to a computational approach for deriving rates of metabolic pathways through a complex network. Typically, a set of input parameters, including the rate of biomass assimilation, nutrient use, and isotopic transfer from precursor to product, is used to constrain possible fluxes throughout the network. Then, a set of equations is used to define the fluxes that, statistically speaking, provide the most plausible fit of the data. Models are tailored to include a broader or more focused swath of the overall metabolome, depending on the hypothesis to be tested and number of input parameters available. Although MFA is a relatively new discipline with evolving opinions about how to report and validate the data, it is evident that understanding fluxes quantitatively will have a significant impact in biomedical research. This is particularly true in disease-oriented research, in which altered metabolic states may not be fully captured by less-quantitative techniques.

Cultured cell models are quite amenable to MFA and, to date, most applications of quantitative flux measurements involve either microorganisms or cultured mammalian cells. MFA was used to describe two fundamentally different biological states in Chinese hamster ovary (CHO) cells (Ahn and Antoniewicz 2011), which are widely used in the pharmaceutical industry to produce recombinant proteins for human therapeutics. A firm understanding of their metabolism may help optimize their performance in such processes. CHO cells were studied during exponential growth and during an early phase of reduced growth rates. $^{13}$C-glucose and $^{13}$C-glutamine were used in parallel experiments to analyze the metabolic network in both growth states. This analysis produced a detailed picture of the metabolic alterations that accompanied the reduced need for biomass in the cells whose growth rates had plateaued. Importantly, some of these changes could not have been predicted based on a priori knowledge about CHO cell biology. This underscores the potential of MFA as a tool to discover new metabolic pathways or modes of metabolic regulation.

Quantitative assessment of metabolic flux can also be achieved in vivo. For the most part, efforts to quantify metabolic activity in humans have focused on a small number of pathways rather than seeking to provide an inclusive description of a broad metabolic network; in this sense, these efforts are better characterized as focused flux measurements than the broader descriptions provided by MFA. Nevertheless, stable isotope tracing in humans can quantify pathways with established relevance to disease states. For example, nonalcoholic fatty liver disease (NAFLD) is a common condition characterized by increased hepatic triglyceride content in the absence of chronic alcohol exposure. It is associated with obesity and insulin resistance, both of which are major causes of mortality in the developed world. Although abundant evidence pointed to an altered state of mitochondrial metabolism in NAFLD, neither the nature of these alterations nor their effects on the rate of mitochondrial pathways were known. In one study, a combination of $^{13}$C and $^{2}$H tracers was used to assess rates of mitochondrial pathways in human patients with or without excess hepatic fat. Analyzing a specific set of metabolic outputs using a focused set of equations allowed these investigators to conclude that NAFLD was associated with an enhanced rate of oxidative metabolism in the liver (Sunny et al. 2011). This provided new insight into the pathological process of NAFLD, and suggested new ways to prevent the abnormal deposition of hepatic fat.

Finally, it is also possible to quantify specific enzyme-catalyzed metabolic activities in vivo, including direct assessment of such activities in diseased tissues. As outlined above, $^{13}$C is a useful tracer for metabolism studies. It is generally not very useful as a nucleus for imaging in vivo because of its low abundance and unfavorable gyromagnetic ratio. However, "hyperpolarization" of $^{13}$C enables this nucleus to be detected noninvasively by spectroscopy with extremely fast acquisition times.

Hyperpolarization involves the temporary redistribution of the populations of the available energy levels of a $^{13}$C nucleus into a nonequilibrium state. This produces a massive gain—more than 10,000-fold, in many cases—in magnetic resonance signal (Ardenkjaer-Larsen et al. 2003). This gain in signal disappears rapidly, usually within a couple of minutes. However, even a temporary enhancement of that magnitude is sufficient to detect both a $^{13}$C-enriched substrate and some downstream metabolites with a temporal resolution of seconds. Although these hyperpolarization studies do not provide a detailed quantitative view of the metabolic network, they do allow a minimally invasive assessment of specific metabolic activities related to the disease state and provide a semiquantitative comparison of rates between normal and diseased tissue. For example, as discussed elsewhere in this book, cancer cells often undergo a metabolic transition known as the Warburg effect (see Chapter 3), wherein a large fraction of glucose is metabolized to lactate. In some human tumors, enhanced activity of lactate dehydrogenase primes the tissue to convert pyruvate to lactate, and this is associated with large lactate pools in the tumor. In vivo detection of hyperpolarized $^{13}$C is being developed as an approach to differentiate between benign and malignant lesions in the prostate. Patients with suspected prostate cancer were injected with hyperpolarized [1-$^{13}$C]pyruvate while the NMR spectra of the prostate tissue were being acquired. In some patients, enhanced exchanges between pyruvate and lactate could be observed in malignant prostate tissue, providing a novel, dynamic view of tumor lactate dehydrogenase activity (Nelson et al. 2013).

## FUTURE DIRECTIONS OF METABOLOMICS

Metabolism is highly complex and highly relevant to human disease. Assessments of metabolism in biomedical research may ask: (1) Which metabolites are present? (2) Which pathways are active? (3) How processive is each pathway? In the ideal situation, all three dimensions are assessed, and then comparisons are made between healthy and diseased tissues. NMR, MS, and, in particular, the application of stable isotopes to biological systems, have greatly accelerated the pace of discovery in metabolism research and provided a wealth of new insights into how altered metabolism affects human health.

Exciting challenges lie ahead. Increased use of isotope tracers in human subjects will likely have significant impact on disease research in the next decade. Improved technologies in MS should make it possible to profile ever more metabolites in ever-smaller biological samples, enabling progressively more detailed views of metabolism in specific cellular populations. It is also rapidly becoming feasible to analyze the localization of metabolites and metabolic activities within intact tissues, lending a topographical dimension to metabolism that will ultimately make it possible to understand on yet another important level.

# REFERENCES

Ahn WS, Antoniewicz MR. 2011. Metabolic flux analysis of CHO cells at growth and non-growth phases using isotopic tracers and mass spectrometry. *Metab Eng* **13:** 598–609.

Andronesi OC, Kim GS, Gerstner E, Batchelor T, Tzika AA, Fantin VR, Vander Heiden MG, Sorensen AG. 2012. Detection of 2-hydroxyglutarate in IDH-mutated glioma patients by in vivo spectral-editing and 2D correlation magnetic resonance spectroscopy. *Sci Transl Med* **4:** 116ra4.

Ardenkjaer-Larsen JH, Fridlund B, Gram A, Hansson G, Hansson L, Lerche MH, Servin R, Thaning M, Golman K. 2003. Increase in signal-to-noise ratio of >10,000 times in liquid-state NMR. *Proc Natl Acad Sci* **100:** 10158–10163.

Cheng T, Sudderth J, Yang C, Mullen AR, Jin ES, Mates JM, DeBerardinis RJ. 2011. Pyruvate carboxylase is required for glutamine-independent growth of tumor cells. *Proc Natl Acad Sci* **108:** 8674–8679.

Choi C, Ganji SK, DeBerardinis RJ, Hatanpaa KJ, Rakheja D, Kovacs Z, Yang XL, Mashimo T, Raisanen JM, Marin-Valencia I, et al. 2012. 2-hydroxyglutarate detection by magnetic resonance spectroscopy in IDH-mutated patients with gliomas. *Nat Med* **18:** 624–629.

Dang L, White DW, Gross S, Bennett BD, Bittinger MA, Driggers EM, Fantin VR, Jang HG, Jin S, Keenan MC, et al. 2009. Cancer-associated IDH1 mutations produce 2-hydroxyglutarate. *Nature* **462:** 739–744.

Fan TW, Lane AN, Higashi RM, Farag MA, Gao H, Bousamra M, Miller DM. 2009. Altered regulation of metabolic pathways in human lung cancer discerned by (13)C stable isotope-resolved metabolomics (SIRM). *Mol Cancer* **8:** 41.

Fan J, Ye J, Kamphorst JJ, Shlomi T, Thompson CB, Rabinowitz JD. 2014. Quantitative flux analysis reveals folate-dependent NADPH production. *Nature* **510:** 298–302.

Nelson SJ, Kurhanewicz J, Vigneron DB, Larson PE, Harzstark AL, Ferrone M, van Criekinge M, Chang JW, Bok R, Park I, et al. 2013. Metabolic imaging of patients with prostate cancer using hyperpolarized [1-$^{13}$C]pyruvate. *Sci Transl Med* **5:** 198ra108.

Pope WB, Prins RM, Albert Thomas M, Nagarajan R, Yen KE, Bittinger MA, Salamon N, Chou AP, Yong WH, Soto H, et al. 2012. Non-invasive detection of 2-hydroxyglutarate and other metabolites in IDH1 mutant glioma patients using magnetic resonance spectroscopy. *J Neurooncol* **107:** 197–205.

Sunny NE, Parks EJ, Browning JD, Burgess SC. 2011. Excessive hepatic mitochondrial TCA cycle and gluconeogenesis in humans with nonalcoholic fatty liver disease. *Cell Metab* **14:** 804–810.

Wang TJ, Larson MG, Vasan RS, Cheng S, Rhee EP, McCabe E, Lewis GD, Fox CS, Jacques PF, Fernandez C, et al. 2011. Metabolite profiles and the risk of developing diabetes. *Nat Med* **17:** 448–453.

# ADDITIONAL READING

DeBerardinis RJ, Thompson CB. 2012. Cellular metabolism and disease: What do metabolic outliers teach us? *Cell* **148:** 1132–1144.

Lutz NW, Sweedler JV, Wevers RA. 2013. *Methodologies for metabolomics: Experimental strategies and techniques.* Cambridge University Press, New York.

# Index

Page references followed by b denote boxes; those followed by f denote figures; those followed by t denote tables.

233